AGRICULTURE ISSUES AND POLICIES

SORGHUM

CULTIVATION, VARIETIES AND USES

AGRICULTURE ISSUES AND POLICIES

Additional books in this series can be found on Nova's website under the Series tab.

Additional E-books in this series can be found on Nova's website under the E-books tab.

AGRICULTURE ISSUES AND POLICIES

SORGHUM

CULTIVATION, VARIETIES AND USES

TOMÁS D. PEREIRA
EDITOR

Nova Science Publishers, Inc.
New York

Copyright © 2011 by Nova Science Publishers, Inc.

All rights reserved. No part of this book may be reproduced, stored in a retrieval system or transmitted in any form or by any means: electronic, electrostatic, magnetic, tape, mechanical photocopying, recording or otherwise without the written permission of the Publisher.

For permission to use material from this book please contact us:
Telephone 631-231-7269; Fax 631-231-8175
Web Site: http://www.novapublishers.com

NOTICE TO THE READER

The Publisher has taken reasonable care in the preparation of this book, but makes no expressed or implied warranty of any kind and assumes no responsibility for any errors or omissions. No liability is assumed for incidental or consequential damages in connection with or arising out of information contained in this book. The Publisher shall not be liable for any special, consequential, or exemplary damages resulting, in whole or in part, from the readers' use of, or reliance upon, this material. Any parts of this book based on government reports are so indicated and copyright is claimed for those parts to the extent applicable to compilations of such works.

Independent verification should be sought for any data, advice or recommendations contained in this book. In addition, no responsibility is assumed by the publisher for any injury and/or damage to persons or property arising from any methods, products, instructions, ideas or otherwise contained in this publication.

This publication is designed to provide accurate and authoritative information with regard to the subject matter covered herein. It is sold with the clear understanding that the Publisher is not engaged in rendering legal or any other professional services. If legal or any other expert assistance is required, the services of a competent person should be sought. FROM A DECLARATION OF PARTICIPANTS JOINTLY ADOPTED BY A COMMITTEE OF THE AMERICAN BAR ASSOCIATION AND A COMMITTEE OF PUBLISHERS.

Additional color graphics may be available in the e-book version of this book.

Library of Congress Cataloging-in-Publication Data

Sorghum : cultivation, varieties and uses / editor: Tomas D. Pereira.
    p. cm.
  Includes index.
  ISBN 978-1-61209-688-9 (hardcover)
 1. Sorghum. 2. Sorghum--Varieties. 3. Sorghum--Utilization. I. Pereira, Tomas D.
  SB191.S7S6643 2011
  633.1'74--dc23
                            2011028635

Published by Nova Science Publishers, Inc. † New York

Contents

Preface		vii
Chapter 1	Uses of Sorghum and Value Addition *Satish D. Shewale and Aniruddha B. Pandit*	1
Chapter 2	Bio-fuel Agro-industrial Production System in Sweet Sorghum *Ramesh C. Ray and Shuvashish Behera*	45
Chapter 3	Sweet Sorghum as an Energy Crop *Yanna Liang*	65
Chapter 4	Chemical Composition of Forage Sorghum and Factors Responsible for Increasing Animal Production *Sandeep Kumar, Kaushalya Gupta and U. N. Joshi*	83
Chapter 5	Dual-purpose Sorghum for Grain and Bioenergy Production for Semi-arid Areas of Southern Africa *Itai Makanda, Pangirayi Tongoona[1] and John Derera*	93
Chapter 6	Tolerance to Aluminum and Nutrient Stress in Sorghum Grown under Simulated Conditions of Tropical Acid Soils *M. Shahadat Hossain Khan, Afrin Akhter, Tadao Wagatsuma, Hiroaki Egashira, Idupulapati M. Rao, S. Ishikawa*	109
Chapter 7	Biological Activities of Sorghum Extract and its Effect on Antibiotic Resistance *Silvia Mošovská, Lucia Birošová, Ľubomír Valík*	129
Chapter 8	Bioenergy Production from Sorghum *Ana Saballos*	141
Index		171

PREFACE

Sorghum is an important world crop, used for food, fodder, the production of alcoholic beverages, as well as biofuels. This new book presents current research in the study of the cultivation, varieties and uses of sorghum. Topics discussed include sweet sorghum as an energy crop; tolerance to aluminum and nutrient stress in sorghum cultivation and the biological activities of sorghum extract and its effect on antibiotic resistance.

Chapter 1 – Sorghum (*Sorghum bicolor* L. Moench) is an important drought resistant cereal crop and the fifth largest produced cereal in the world after wheat, rice, barley and maize. Sorghum is valued because of its ability to produce in tropical and arid regions of the world, where it is difficult to grow any other cereal, and also, because of its relatively short growing season requirement, thus its suitability for double cropping and crop rotation systems. Around 10 to 30% of the produced grains gets wasted or damaged. Damage occurs to grains due to breakage, infection by insect and fungus. Disposal of such damaged grains is a major problem from an environmental point of view and it is also wastage of GDP. Utilization of the sorghum can be classified mainly into two categories viz. Human Food, Animal Food and industrial use. Sorghum acts as a principal source of energy, protein, vitamins and minerals for millions of the poorest people living in drought regions, who cultivate sorghum for consumption at home and in certain cases for feeding their cattle. It is consumed as whole grain or processed into flour, from which traditional meals are prepared. The main industries using sorghum are the animal feed sector, alcohol distilleries, and starch industries. It also serves as an important source of cattle feed and fodder. It is grown by United States, Australia and other developed countries for animal feed. Sorghum grows comparatively quicker and gives not only good yields of grain but also very large quantities of fodder. Since, sorghum is rich in starch content (around 60 – 77 %), it is used in production of bio-industrial products like bioethanol, Glucose and it also serves as source material for isolation of starch in scarcity of maize. Malting and brewing industries also utilize sorghum to produce lager and stout. Starch industry utilizes isolated starch (normally corn starch) to produce dextrose and maltose based products. This is particularly done due to achievable high (> 97%) conversion with isolated starch as a starting material. But, isolation yield of starch from corn is around 50-70 % i.e. rest part (30-50%) gets wasted or does not fetch respectable market price. Grain flour (i.e. unisolated starch) of healthy or damaged grains can be directly used for production of dextrose and maltose based products. Use of healthy grains is obviously uneconomical because they have good market demand and obviously high cost. Damaged grains are not of edible quality and are available at much lower prices. Hence, it would be more economical raw material for value addition through production of glucose and maltose based products. However, bottleneck in the use of grain flour from damaged grains is low conversions of starch i.e. 80 %. Ultrasound technology has been successfully integrated in the state of art production of dextrose hydrolysate and very high maltose syrup. With integration of ultrasound technology, around 10 % improvement in the conversions of starch has been observed using two different qualities of sorghum grains viz. 1. healthy sorghum, and 2. blackened sorghum as a starting material. Reaction Kinetics, yields and the economics of the process has been discussed.

Chapter 2 – The consumption of bio-ethanol as bio-fuel may reduce greenhouse gases and gasoline imports. Also it can be replaced with lead or MTBE (Methyl tert-butyl ether) that are air or underground water pollutants, respectively. Plants are the best choice for meeting the projected bio-fuel demands. Recent studies have revealed that sweet sorghum can be used as a feedstock for bio-ethanol and bio-diesel production under hot and dry climatic conditions because it has higher tolerance to salt and drought compared to sugarcane, sugar beet, maize and cassava that are currently used for bio-fuel production in the world. In addition, high carbohydrate contents of sweet sorghum stalk are similar to sugarcane but its water and fertilizer requirements are much lower than sugarcane. The recent developments such as solid state fermentation, simultaneous saccharification and fermentation technology, use of mixed culture, recombinant yeast, and immobilized techniques, and the factors such as pH, agitation, particle stuffing rate, etc. in bio-fuel production from sweet sorghum have been discussed in this chapter.

Chapter 3 – As a potential energy crop, sweet sorghum has attracted considerable attention during recent years. Three kinds of biofuels, ethanol, lipid, and hydrogen can be produced from sweet sorghum through different processes. Whole stalk of sweet sorghum, juice, and bagasse have been explored to generate these biofuels to certain extent. Based on related publications, the best technologies for different biofuel production processes are identified for further investigation. Logistic issues, such as sorghum harvesting and storage are also discussed in this chapter. Energy and economic analyses are described without too many details considering the regionally specific features. In summary, every part of sweet sorghum can be utilized for biofuel production. The cost-benefit and life cycle analysis, however, needs to be investigated thoroughly to ensure that plantation of sweet sorghum for bioenergy purpose is sustainable, economical, and scalable.

Chapter 4 – With the increasing cattle population, the gap between demand and supply of fodder is also increasing, which can be reduced through improvement of forage crops, especially forage sorghum. Forage sorghum (*Sorghum bicolor* L. Moench) is grown mainly for fodder purpose during summer and *kharif* seasons in the northern states of India. Kamalak et al., (2005) reported that protein upto a minimum level (7 %) is essential to maintain rumen micro-flora for proper digestion as well as intake of the fodder. The productivity of the livestock depends upon good quality fodders. Animal performance mainly depends on three factors namely intake, digestibility and utilization.

Chapter 5 – Sorghum (*Sorghum bicolour* L Moench) is one of the most important cereal crops in sub-Saharan Africa (SSA). It plays a major food security role in the semi-arid areas in southern Africa where small-scale and resource-poor farmers reside. Apart from food, sorghum has high potential as an industrial crop in the biofuel industry. Sorghum varieties have been traditionally developed for grain, especially in SSA, whereas specialist varieties for fodder or stem sugar have been bred in developed countries. What could be most appealing to small-scale farmers in SSA are dual-purpose varieties that combine high grain yield and stem sugar content. When developed, such varieties would be beneficial to these resource-poor farmers by providing adequate grain for food and sugar rich stalks for commercial production of bioethanol (used as biofuel) with possible multiplier effects. However, a survey of the literature indicates clearly that breeding effort has been limited to specialised cultivars alone. Information on the inheritance, combining ability, gene action and relationships between stem sugar and grain yield traits in breeding dual-purpose sorghum source germplasm is limited. This information is crucial for devising appropriate strategies for developing dual-purpose sorghum varieties. If grain yield and stem sugar are mutually exclusive, then breeding for the two traits in a single cultivar will be a challenging task. Results from a few studies have indicated a weak and non-significant association between these traits. These findings pertain to the populations and test environments which were sampled; therefore, a careful selection of base populations is key to achieving progress in breeding such cultivars in southern Africa. Further, there is also lack of information about perceptions and views of stakeholders on the potential of dual-purpose sorghum production, utilisation and the general value chain. Therefore, this review is an attempt at elucidating the information necessary to establish a viable breeding programme for dual-purpose sorghum cultivars. It covers the utilisation of sorghum for food and biofuel, benefits of the cultivars to the farmers, stakeholders' views on the technology; the genetics of grain yield and stem sugar content; and the relationship between the two traits. Implications of the findings and suggestions for the way forward are also discussed.

Chapter 6 – Acid soils occur mainly in two global belts: the northern belt, with cold and humid conditions, and the southern tropical belt, with warmer humid conditions. Although aluminum (Al) is usually regarded as the determining factor for growth of many crop plants in acid soils, nutrient deficiencies are also a major predicament in tropical acid soils. Sorghum (*Sorghum bicolor* Moench [L]) is generally an Al-sensitive crop species. While screening for the differential Al tolerance among sorghum cultivars, we found that lower level of Al in culture solutions increased the coefficient of variation in sorghum cultivars than in the case of maize cultivars. Contrary to the common agreement on the greater contribution of Al tolerance for improved adaptation to acid soils, appropriate strategy is needed for sorghum production in tropical acid soils. This observation is based on the comprehensive but preliminary evaluation under long-term conditions of combined stress conditions with varying concentrations of Al and nutrients in solution culture simulating the nutrient stress of tropical acid soils. Limited research has been carried out considering these two factors, i. e., concentrations of Al and nutrients, concurrently. In sorghum, a greater tolerance under combined stress conditions was associated with a higher shoot K concentration. Although Al tolerance is considered as an important strategy for sorghum production in tropical acid soils, plant nutritional characteristics linked to low nutrient tolerance can be the primary factor for better growth of maize cultivars that are tolerant to Al.

Chapter 7 – Sorghum is an ancient crop belonging to the gluten-free cereals that are known as pseudocereals. It is an important food in semi-arid tropics of the world including Africa or Asia because sorghum as other pseudocereals is extensively drought tolerant and well-adapted to weather extremes. It is quite widely used. It is applied as a forage crop or as a raw material for bio ethanol and other industrial products. At present, the interest in sorghum is growing considerably as a potential crop in food industries. Since sorghum does not contain gluten it is suitable for people with celiac diseases. Seeds of sorghum are also a good source of various phytochemicals such as phenolic compounds including phenolic acids, flavonoids and condensed tannins which have potential biological activities. The objective of this study was to investigate antioxidant and antimicrobial activity of sorghum extract. The ability of extract to quench free radicals was measured by spectrophotometric ABTS (IC_{50} 0,811 mg/ml), DPPH (IC_{50} 1,645 mg/ml) and FRAP (IC_{50} 0,569 mg/ml) methods. Extract inhibited growth of fungi *Aspergillus flavus* and *Alternaria alternata* and showed also antibacterial effect on gram-positive (*Bacillus subtilis, Staphylococcus aureus, Staphylococcus epidermidis*) and gram-negative (*Salmonella typhimurium, Pseudomonas aeruginosa, Escherichia coli, Enterobacter sakazakii*) bacteria. In the last years, the frequency of antimicrobial resistant infection has increased and thus antibacterial resistance has become a serious problem. The various experimental studies suggested that antimutagens could have an important function in the war against microbial resistance. Whereas our previous results indicated potential antimutagenic activity of tested extract, effect of sorghum on the development of antibiotic resistance was also studied.

Chapter 8 – Sorghum bicolor is the 5th most cultivated cereal crop in the world. It has been used by humans for over 5,000 years for food, feed and fodder and it is a staple crop in arid, nutrient-poor regions of the world due to its tolerance to harsh growing conditions. Its resilient nature allows its cultivation with fewer inputs than other crops such maize, wheat and rice. Because of its resiliance and high biomass yield potential, there is a growing interest in sorghum cultivation as a bioenergy crop. Three production streams for biofuels and bio-based products can be obtained from the species: Stanch from the grain, sugar from the sweet stalks and lignocellulosic biomass from the bagasse or stover. Current sorghum varieties and hybrids were not bred for bioenergy applications, and consequently, they do not take full advantage of the genetic variation present in the species. Efforts to breed dedicated bioenergy varieties are underway, with focus in incorporating traits such as high biomass, improved sugar production, and increased stover and starch saccharification potential. This review will cover the current and proposed methodologies for bioenergy production from sorghum, and the research underway to improve the feedstock quality of the species.

In: Sorghum: Cultivation, Varieties and Uses
Editor: Tomás D. Pereira

ISBN 978-1-61209-688-9
© 2011 Nova Science Publishers, Inc.

Chapter 1

USES OF SORGHUM AND VALUE ADDITION

Satish D. Shewale[1] and Aniruddha B. Pandit[2]
[1]Process engineering and scale up department,
Pharmaceutical intermediate division, Atul Ltd, India
[2]University Institute of Chemical Technology,
UGC, Research Scientist 'C', University of Mumbai, India

ABSTRACT

Sorghum (*Sorghum bicolor* L. Moench) is an important drought resistant cereal crop and fifth largest produced cereal in the world after wheat, rice, barley and maize. Sorghum is valued because of its ability to produce in tropical and arid regions of the world, where it is difficult to grow any other cereal, and also, because of its relatively short growing season requirement, thus its suitability for double cropping and crop rotation systems. Around 10 to 30% of the produced grains gets wasted or damaged. Damage occurs to grains due to breakage, infection by insect and fungus. Disposal of such damaged grains is a major problem from environmental point of view and it is also wastage of GDP.

Utilization of the sorghum can be classified mainly into two categories viz. Human Food, Animal Food and industrial use. Sorghum acts as a principal source of energy, protein, vitamins and minerals for millions of the poorest people living in drought regions, who cultivate sorghum for consumption at home and in certain cases for feeding their cattle. It is consumed as whole grain or processed into flour, from which traditional meals are prepared. The main industries using sorghum are the animal feed sector, alcohol distilleries, and starch industries. It also serves as an important source of cattle feed and fodder. It is grown by United States, Australia and other developed countries for animal feed. Sorghum grows comparatively quicker and gives not only good yields of grain but also very large quantities of fodder. Since, sorghum is rich in starch content (around 60 – 77%), it is used in production of bio-industrial product like bioethanol, Glucose and it also serves as source material for isolation of starch in scarcity of maize. Malting and brewing industries also utilize sorghum to produce lager and stout.

Starch industry utilizes isolated starch (normally corn starch) to produce dextrose and maltose based products. This is particularly done due to achievable high (> 97%) conversion with isolated starch as a starting material. But, isolation yield of starch from corn is around 50-70% i.e. rest part (30-50%) gets wasted or does not fetch respectable market price. Grain flour (i.e. unisolated starch) of healthy or damaged grains can be directly used for production of dextrose and maltose based products. Use of healthy grains is obviously uneconomical because they have good market demand and obviously high cost. Damaged grains are not of edible quality and are available at much lower prices. Hence, it would be more economical raw material for value addition through production of glucose and maltose based products. However, bottleneck in the use of grain flour from damaged grains is low conversions of starch i.e. 80%. Ultrasound technology has been successfully integrated in the state of art in production of dextrose hydrolysate and very high maltose syrup. With integration of ultrasound technology, around 10% improvement in the conversions of starch has been observed using two different qualities of sorghum grains viz. 1. healthy sorghum, and 2. blackened sorghum as a starting material. Reaction Kinetics, yields and the economics of the process has been discussed.

1. INTRODUCTION

Sorghum (*Sorghum bicolor* L. Moench) is an important drought resistant cereal crop and fifth largest produced cereal in the world after wheat, rice, barley and maize. Sorghum is termed as "Nature-cared crop" because it has strong resistance to harsh environments such as dry weather and high temperature in comparison to other crops, it is usually grown as a low-level chemical treatment crop with limited use of pesticides and it has a potential to adapt itself to the given natural environment. Better drought tolerance of sorghum than most other grain crops can be attributed to: leaves rolling ability in moisture stress; thin wax layer covering leaves; well developed and finely branched root system; limited transpiration due to smaller leaf area per plant; tolerance to short water logging; ability to remain in a virtually dormant stage during dry periods and resume growth as soon as conditions become favorable. Sorghum is valued because of its ability to grow in areas with marginal rainfall and high temperatures (i.e. semi arid tropics and sub tropical regions of the world), where it is difficult to grow any other cereal, and also because of its relatively short growing season requirement, thus its suitability for double cropping and crop rotation systems. Sorghum is key crop in providing food security to people in semi arid tropics and sub tropical regions of the world.

Sorghum belongs to the Grass family. It is broadly classified into four types viz. 1. Grain sorghum (Food, feed and industrial use) 2. Sweet sorghum (for sweetner syrup) 3. Grass sorghum (feed and forage use) 4. Broomcorn sorghum (used in making brooms).

Sorghum is a staple food for about 300 million people worldwide. The seed or caryopsis of sorghum provides a major source of calories and protein for millions of people in Africa and Asia. However, the demand for sorghum for human consumption is decreasing with change in the way of living due to increased urbanization, increased per capita income of the population, and easy availability of other preferred cereals in sufficient quantities at affordable prices. Hence, sorghum also serves as a source of feed for cattle and other livestock in scarcity of maize, but at lower prices. Sorghum grows comparatively quicker and gives not only good yields of grain but also very large quantities of fodder. It is grown by United States, Australia and other developed countries for animal feed. Since, sorghum is rich in starch content (around 60 – 77%), it serves as source material for isolation of starch in scarcity of maize. Starch has applications in textile, food and pharmaceutical industries. Starch can be processed to produce dextrose, maltose based products, high fructose syrup, etc. Sorghum grain or isolated starch is used in production of bioethanol through fermentation. Malting and brewing industries also utilize sorghum to produce lager and stout.

Around 10 – 30% of the production gets wasted due to damage, and inadequate transport and storage facilities. Damage occurs to grains due to breakage, inadequate storage facility, infection by insect and fungus. Disposal of such damaged grains is a major problem from an environmental point of view and it is also wastage of GDP. Damaged grains are not of edible quality and are available at much lower prices. Hence, it would be more economical raw material for value addition through production of bioethanol, glucose and maltose based products. Also, industrial use of healthy grains would be uneconomical because they have good market demand and obviously high cost.

2. ORIGIN AND GEOGRAPHICAL DISTRIBUTION OF SORGHUM

Sorghum is believed to be originated in equatorial Africa, where a large variability in wild and cultivated species is still found today. It was probably domesticated in Ethiopia between 5,000 and 7,000 years ago. From there, it was distributed along trade and shipping routes around the African continent, and through the Middle East to India at least 3,000 years ago. It is believed that from India it was carried to China along the silk route and through coastal shipping to South-East Asia. Sorghum was first taken to America through the slave trade from West Africa. It was introduced into the United States for commercial cultivation from North Africa, South Africa and India at the end of the 19th century and subsequently spread to South America and Australia. It is now widely cultivated in dry areas of Africa, Asia, the Americas, Europe and Australia between latitudes of up to 50°N in North America and Russia and 40°S in Argentina. (Kimber et al., 2000; Balole and Legwaila, 2005)

Sorghum is distributed throughout the tropical, semi-tropical, arid and semi arid regions of the world. Sorghum is also found in temperate regions and at altitudes of up to 2,300 meters in the tropics. It has a potential to compete effectively with crops like maize under good environmental and management conditions. It is one of the most widely grown dry land food grains in world. It does well even in low rainfall areas.

In Africa, a major growing area runs across West Africa south of the Sahara, through Sudan, Ethiopia and Somalia. It is grown in upper Egypt and Uganda, Kenya, Tanzania, Burundi, and Zambia. It is an important crop in India, Pakistan, Thailand, in central and northern China, Australia, in the dry areas of Argentina and Brazil, Venezuela, U.S.A., France and Italy. Sorghum is called by various names in different places in the world. Sorghum is known by various names in Africa: as guinea-corn, dawa or sorgho in West Africa, durra in the Sudan, mshelia in Ethiopia and Eritrea, mtama in East Africa, kaffircorn in South Africa and mabele or amabele in several countries in Southern Africa. It is called jowar in India, kaolian in China and milo in Spain. In the Indian subcontinent, it is known as jowar, jwari (Maharashtra), jonna (Andhra Pradesh), cholam (Tamil Nadu) and jola (Karnataka).

3. TAXONOMY

Pliny (ca. 60 to 70 A. D.) was the first to give a written description of sorghum and after that there was hardly a mention of it until the sixteenth century. Moench in 1794 established the genus Sorghum and brought the sorghums under the name Sorghum bicolor. Harlan and de Wet (1972) developed a simplified classification that has real practical utility for sorghum workers. Sorghum (L.) Moench comprises about 20-30 species. Sorghum Bicolor (L.) Moench is the primarily cultivated species. Other perennial species being Sorghum almum (Columbus grass), Sorghum halepense (Johnson grass) and Sorghum propinquum. Subspecies of Sorghum Bicolor (L.) Moench are arundinaceum, bicolor and drummondii (Dahlberg, 2000).

Sorghum bicolor (L.) Moench subspecies bicolor i.e. grain sorghum contains all of the cultivated sorghum and is sub classified into different races on the basis of grain shape, glume shape, and panicle shape. Five basic races are Bicolor, Guinea, kafir, Caudatum, and Durra. Different intermediate races are guinea-bicolor, caudatum-bicolor, kafir-bicolor, durra-bicolor, guinea-caudatum, guinea-kafir, guinea-durra, kafir-caudatum, durra-caudatum, kafir-durra. Hybrid races exhibit various combinations and intermediate forms of the characteristics of the 5 basic races. Durra-bicolor is found mainly in Ethiopia, Yemen and India, guinea-caudatum is a major sorghum grown in Nigeria and Sudan, and guinea-kafir is grown in East Africa and India. Kafir-caudatum is widely grown in the United States and almost all of the modern North American hybrid grain cultivars are of this type. Guinea-caudatum with yellow endosperm and large seed size is used in breeding programmes in the United States. The species Sorghum bicolor covers a wide range of varieties, from white and yellow to brown, red and almost black. Classification and characterization of sorghum is given in detail in Dahlberg (2000).

Taxonomical hierarchy of Sorghum bicolor (L.) Moench (www.itis.gov) is as follows:

Kingdom Plantae – Plants
Subkingdom Tracheobionta – Vascular plants
Sperdivision Spermatophyta –Seed plants
Division Magnoliophyta – Flowering plants
Class Liliopsida – Monocotyledons
Subclass Commelinidae
Order Cyperales
Family Poaceae – graminées, grass family
Genus Sorghum Moench – sorghum
Species Sorghum bicolor (L.) Moench – black amber, broomcorn, chicken corn, shatter cane, shattercane, sorghum, wild cane
Subspecies (ssp.)
Sorghum bicolor ssp. Arundinaceum – Common wild sorghum
Sorghum bicolor (L.) Moench ssp. bicolor – grain sorghum
Sorghum bicolor ssp. drummondii – Sudangrass
Synonyms for Sorghum bicolor – Black amber, broom-corn, broomcorn, chicken corn, common wild sorghum, Drummond broomcorn, durra, Egyptian millet, feterita, forage

sorghum, great millet, guinea corn, jowar, Kaffir-corn, Kaffircorn, milo, shallu, shatter cane, shattercane, sorghum, Sudan Grass, sweet sorghum and wild cane.

Synonyms for Sorghum bicolor ssp. Bicolor – Holcus bicolor, Holcus sorghum, Sorghum bicolor var. caffrorum, Sorghum caffrorum, Sorghum cernuum, Sorghum dochna, Sorghum dochna var. technicum, Sorghum drummondii, Sorghum durra, Sorghum saccharatum, Sorghum subglabrescens, Sorghum vulgare, Sorghum vulgare var. caffrorum, Sorghum vulgare var. durr, Sorghum vulgare var. roxburghii, Sorghum vulgare var. saccharatum and Sorghum vulgare var. technicum.

Sorghum is genetically diverse. The world sorghum germplasms are deposited at the International Crop Research Institute for the Semi-Arid Tropics (ICRISAT, Patancheru) in India. ICRISAT holds about 36,000 germplasm accessions of this crop. The varieties are distinguished on the basis of morphological traits, differences in isoenzyme patterns and DNA polymorphism.

4. PRODUCTION, CULTIVATION AREA AND YIELD OF SORGHUM

Sorghum is most widely grown in the semi-arid tropics, where water availability is limited and is frequently subjected to drought. Sorghum cultivation is distributed throughout the world in about 100 countries (Figure 1). In Asia, it is grown in China, India, Korea, Pakistan, Thailand and Yemen. Australia and U.S.A. grow the crop too. Here one thing should be remembered: that these developed countries cultivate sorghum for animal feed, whereas, developing countries in Asia and Africa cultivate it for use as human food. In Southern and Eastern Africa, the sorghum-growing countries are Botswana, Eritrea, Kenya, Lesotho, Madagascar, Malawi, Mozambique, Namibia, Somalia, South Africa, Swaziland, Tanzania, Zambia and Zimbabwe. In West and Central Africa, the crop is grown in Benin, Burkina Faso, Burundi, Cameroon, Central African Republic, Chad, Egypt, Gambia, Ghana, Guinea, Guinea-Bissau, Ivory Coast, Mali, Mauritania, Morocco, Niger, Nigeria, Rwanda, Senegal, Sierra Leone, Sudan, Togo, Tunisia and Uganda. In Latin America, the sorghum-growing countries are Argentina, Brazil, Colombia, El Salvador, Guatemala, Haiti, Honduras, Mexico, Nicaragua, Peru, Uruguay and Venezuela. In Europe, it is grown in France, Italy, Spain, Albania and Romania. (Deb et al., 2004)

Production of sorghum in 2009-2010 in the world was 60 Million Metric Tons (www.fas.usda.gov). Until 2008-09 the United States was the largest sorghum producing country followed by Nigeria, India, Mexico and Sudan. In 2009-10 Nigeria (19.4 %) became largest sorghum producing country followed by the United States (16.4 %), India (11.8 %), Mexico (9.9), Argentina (6.4 %) and Sudan (4.4%) (Figure 2).

India had the largest area under sorghum cultivation of 7.7 million ha in 2008-09. The second largest sorghum cultivating country was Nigeria, followed by Sudan, U.S.A. and Mexico. Around 90% of the world's sorghum cultivation area lies in the developing countries, mainly in Africa and Asia.

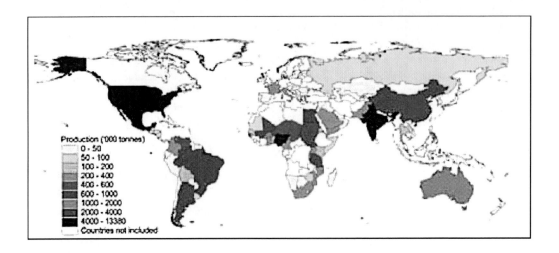

Figure 1. Distribution of sorghum production, 1999-2001. (Deb *et al.*, 2004).

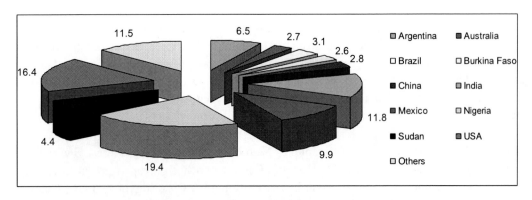

Figure 2. Percent production of sorghum country-wise with world production of 60 Million Tonnes in 2009-2010. (Source of data: www.fas.usda.gov).

Countrywide trend in production, cultivation area and yield of sorghum from 1970 to 2009 is given in Table 1. The United States, which was the highest sorghum producing country until 2009, showed consistent decline in sorghum production since 1970. Nigeria is showing consistent increase in sorghum production and in 2009-10 it overtook the United States and became the largest sorghum producing country in world. Nigeria shows 274% increase and the United States shows 56% decrease in sorghum production in 2009-10 as compared to that in the 1970s. The United States, China and Argentina show consistent decline in sorghum production since 1970 in accordance with consistent decline in sorghum cultivation area. India shows increase in sorghum cultivation area and production from 1970 to 1983; however since then it shows consistent slow downfall in both. Almost all African countries like Nigeria, Sudan, Tanzania, Niger are showing steady increase in sorghum cultivation area and production as well. This also underlines the increasing importance of sorghum on Africa.

Table 1. Countrywide trend in production, cultivation area and yield of sorghum

Country	Average production (' 000 Ton) in year						
	71-73	81-83	91-93	99-2001	2003-04	2008-09	2009/10
Argentina	4140	7935	2626	3159	2200	1660	3850
Australia	1181	1160	915	1810	2009	2690	1600
Brazil	85	224	274	743	2103	1910	1825
Burkina Faso	489	626	1280	1131	1520	1875	1522
China	8680	7343	5151	2948	2865	1837	1650
Egypt	846	623	740	945	900	900	900
Ethiopia					1700	2619	2084
India	7929	11578	10588	8232	6680	7310	6980
Mali	284	452	716	649	n.a.	930	980
Mexico	2799	5286	4582	6092	7300	7067	5850
Niger	200	345	424	501	720	1214	739
Nigeria	3072	3589	4832	7647	9000	11000	11500
Sudan	1527	2300	3323	2441	5190	4192	2630
Tanzania	172	493	619	654	500	700	710
USA	21951	18614	16839	13380	10446	11998	9728
Others				8224	5145	6873	6809
World total				58556	58278	64775	59357

Country	Average cultivation area (' 000 ha) in year						
	71-73	81-83	91-93	99-2001	2003-04	2008-09	2009/10
Argentina	2074	2411	721	690	n. a.	450	n. a.
Australia	629	671	460	602	n. a.	770	n. a.
Brazil	50	117	159	453	n. a.	850	n. a.
Burkina Faso	1038	1073	1417	1302	n. a.	1620	n. a.
China	5072	2704	1368	941	n. a.	490	n. a.
Egypt	205	166	144	163	n. a.	160	n. a.
Ethiopia				1190	n. a.	1550	n. a.
India	16335	16469	12574	10056	n. a.	7700	n. a.
Mali	373	534	875	719	n. a.	n a	n. a.
Mexico	1077	1520	1305	1992	n. a.	1890	n. a.
Niger	531	1075	2315	2286	n. a.	1500	n. a.
Nigeria	4792	2216	4535	6816	n. a.	7400	n. a.
Sudan	1974	3682	5345	4307	n. a.	6400	n. a.
Tanzania	338	500	642	639	n. a.	900	n. a.
USA	6077	5101	4160	3353	n. a.	2940	n. a.
Others				6350		7400	
World total				41859	n. a.	42020	n. a.

Country	Average yield (kg/ha) in year						
	71-73	81-83	91-93	99-2001	2003-04	2008-09	2009-10
Argentina	1996	3291	3642	4578	n. a.	3689	n. a.
Australia	1878	1729	1989	3007	n. a.	3494	n. a.
Brazil	1700	1915	1723	1640	n. a.	2247	n. a.
Burkina Faso	471	583	903	869	n. a.	1157	n. a.
China	1711	2716	3765	3133	n. a.	3749	n. a.
Egypt	4127	3753	5139	5798	n. a.	5625	n. a.
Ethiopia	n a	n. a.	n. a.	n. a.	n. a.	1690	n. a.
India	485	703	842	819	n. a.	949	n. a.
Mali	761	846	818	903	n. a.	n. a.	n. a.
Mexico	2599	3478	3511	3058	n. a.	3739	n. a.
Niger	377	321	183	219	n. a.	809	n. a.
Nigeria	641	1620	1065	1122	n. a.	1486	n. a.
Sudan	774	625	622	567	n. a.	655	n. a.
Tanzania	509	986	964	1023	n. a.	778	n. a.
USA	3612	3649	4048	3990	n. a.	4081	n. a.

Source of data: Deb et al. (2004) and www.fas.usda.gov.

Major sorghum producing countries viz. Nigeria, United States, India, Sudan, Mexico, and Argentina have yields (kg/ha) of 1,486, 4,081, 949, 655, 3,739, and 3,689 kg/ha, respectively (Table 1) in 2008-09. Deb et al., 2004 reported that highest sorghum yields during 1999-2001 were recorded by Israel (12,664 kg/ha), followed by Jordan (11,711 kg/ha), Italy (6,458 kg/ ha) and Algeria (6,400 kg/ha). Here, It should be noted that Israel and Jordan are not major sorghum-growing countries. Maximum yield of 12,664 kg/ha of Israel was around 15 times the sorghum yield of India. Deb et al., 2004 also commented that though Asian and African countries like India and Nigeria had the largest area devoted to sorghum cultivation, countries in West Asia (like Israel and Jordan) and Europe (Italy and France) reaped the highest yields.

In India sorghum is grown in the kharif (rainy season) and rabi (postrainy season). The share of kharif is higher both in terms of area under cultivation and production. The kharif sorghum crop accounts for 55% of the total area under cultivation and 68% of the total production. Maharashtra was the largest sorghum producing state in 2001-2002 with a share of around 50%. In fact, the niche of sorghum production primarily remains in the two states of Maharashtra and Karnataka in India, where area under sorghum production stands at a total of 7 million ha. (Source of data: www.agricoop.nic.in).

Table 2 comprises countrywide data of consumption and import/export of sorghum. United States is the largest exporter of sorghum in the world followed by Argentina and Australia. World trade in sorghum is dominated by U.S.A.. U.S.A. contributes to around 70% of the total export of sorghum in world. United States has been exporting around four – five million metric tonnes of sorghum grains annually, generating an income of around half a billion dollars. Nigeria, India and Mexico are major consumers of sorghum in descending order. Nigeria and India are consuming almost the entire quantity of sorghum that they are producing. Demand of Sorghum in Mexico is large and exceeds significantly their sorghum production, making Mexico the largest importer of sorghum. Japan is the second largest importer of sorghum. Mexico and Japan, together, import around 40% of international imports.

Sorghum is a major input to the Mexican feed industry representing sixty to eighty per cent of mixed feeds that are used in animal feeding. International market prices for sorghum are largely determined by the supply and demand situation in the United States, and export prices are based on the reference sorghum, US Milo No. 2, yellow (Jeyakumar, 2010).

World trade in sorghum is strongly linked to demand for livestock products, dominated by feed requirements and prices in Group II countries. Only 6 percent of world sorghum trade (about 500,000 tonnes per year) is for use as food. This is mainly imported by countries in Africa. Since trade is primarily for animal feed, volumes are very sensitive to sorghum/maize price differentials and can fluctuate considerably (www.FAO.org).

5. GRAIN MORPHOLOGY

Waniska and Rooney (2000) have reviewed and given the grain morphology in detail. Physical composition (i.e. distribution in pericarp, germ, and endosperm) and chemical composition (i.e. starch, protein, fiber, lipid and ash) of sorghum grain is given in Table 3.

Table 2. Consumption and Trade of sorghum

Country	Consumption in year			Export (+) / Import (-) in year		
	2003-04	2008-09	2009-10	2003-04	2008-09	2009-10
Argentina	2000	1100	2300	200	560	1550
Australia	1405	1805	1005	604	885	595
Brazil	1650	1900	1900	453	10	-75
Burkina Faso	1520	1875	1522	0	0	0
China	3000	2000	1900	-135	-163	-250
Egypt	900	950	900	0	-50	0
Ethiopia	1700	2700	2400	0	-81	-316
India	6700	7200	6900	-20	110	80
Japan	1500	1600	1800	-1500	-1600	-1800
Mali	n.a.	930	980	n.a.	0	0
Mexico	9800	8600	9500	-2500	-1533	-3650
Niger	750	1200	850	-30	14	-111
Nigeria	8950	10950	11450	50	50	50
Sudan	4500	4500	3400	690	-308	-770
Others	7507	8075	8102	-7007	-7375	-7392
United States	5638	8319	5853	4808	3679	3875
World Total	57520	64481	61608			

Source of data: www.fas.usda.gov.

Table 3. Composition of sorghum seed in %

	Caryopsis	Endosperm	Germ	Pericarp
Caryopsis	100	84.2	9.4	6.5
Range	–	81.7 – 86.5	8.0 – 10.9	4.3 – 8.7
Protein	11.3	10.5	18.4	6.0
Range	7.3 – 15.6	8.7 – 13.0	17.8 – 19.2	5.2 – 7.6
Distribution	100	80.9	14.9	4.0
Fiber	2.7	–	–	–
Range	1.2 – 6.6	–	–	–
Distribution	100	–	–	–
Lipid	3.4	0.6	28.1	4.9
Range	0.5 – 5.2	0.4 – 0.8	26.9 – 30.6	3.7 – 6.0
Distribution	100	13.2	76.2	10.6
Ash	1.7	0.4	10.4	2.0
Range	1.1 – 2.5	0.3 – 0.4	–	–
Distribution	100	20.6	68.6	10.8
Starch	71.8	82.5	13.4	34.6
Range	55.6 – 75.2	81.3 – 83	–	–
Distribution	100	94.4	1.8	3.8

Source: Waniska and Rooney, 2000.

The caryopsis (seed) of sorghum consists of three distinct anatomical components: pericarp (outer layer), endosperm (storage tissue), and germ (embryo). The outer layer or pericarp originates from the ovary wall and is comprised of three segments viz. epicarp, mesocarp, and endocarp. The epicarp is two or three cell layers thick and consists of rectangular cells that often contain pigmented material. The thickness of the mesocarp, the middle structure, varies from the very thin cellular layer (containing a small amount of starch granules) to 3 or 4 cellular layers containing a large amount of starch granules. Sorghum is the only food grade crop that is reported to contain starch in this anatomical section. The innermost endocarp is composed of cross cells and tube cells.

The endosperm tissue is triploid, resulting from the fusion of a male gamete with two female polar cells. It is composed of the aleurone layer, peripheral, corneous and floury areas. The aleurone is the outer cover and consists of a single layer of rectangular cells adjacent to the testa or tube cells. The cells possess a thick cell wall, large amounts of proteins (protein bodies, enzymes), ash (phytin bodies), and oil (spherosomes). The peripheral area is composed of several layers of dense cells containing more protein and smaller starch granules than the corneous area. Both the peripheral and corneous areas appear translucent, or vitreous, and they affect processing and nutrient digestibility. Waxy sorghums contain larger starch granules and less protein in the peripheral endosperm than regular sorghums.

The corneous and floury endosperm cells are composed of starch granules, protein matrix, protein bodies, and cell walls rich in cellulose, β-glucans, and hemicellulose. Starch granules and protein bodies are embedded in the continuous, protein matrix in the peripheral and corneous areas. The protein bodies are largely circular and 0.4–2.0 µm in diameter. High-lysine cultivars contain fewer and smaller protein bodies than do regular sorghums, and thus contain significantly less alcohol soluble kafirins. The starch granules are polygonal and often contain dents from the protein bodies. Size of starch granules varies from 4 µm to 25 µm, the average being 15 µm. Granules present in the corneous endosperm are smaller and angular whereas those in the floury endosperm are larger and spherical. The opaque, floury endosperm (located near the center of the caryopsis) has a discontinuous protein phase, air voids, and loosely packaged, round, starch granules. The presence of air voids diffracts incoming light giving an opaque or chalky appearance.

The germ is diploid due to the sexual union of one male and one female gamete. The protein of the germ contains high levels of lysine and tryptophan that are excellent in quality.

Sorghum grain consists of carbohydrates, proteins, fibers, lipids etc. Carbohydrates in sorghum are composed of starch, soluble sugar and fiber (pentosans, cellulose, and hemicellulose).

Starch

Starch is most abundant in sorghum grain and is important from the perspective of industrial utilization. Sorghum starch has properties and uses similar to those of maize starch and the procedure for wet milling of sorghum is similar to the one used for maize. Pigments in the sorghum pericarp discolors the starch, yielding a light pink color. Bleaching with $NaClO_2$ or rinsing with NaOH or methanol produces the acceptable color. Normal sorghum starch contains 23 – 30% amylose. Average molecular weights of amylose and amylopectin were 1 to 3×10^5 and 8 to 10×10^6 kD, respectively. Sorghum that has three recessive wx genes produces caryopses that contain mostly amylopectin and are termed as waxy sorghum. Waxy sorghum consists of 1% amylose. Heterowaxy sorghum (inclusive of one or two wx genes) consists of 5 to 19% amylose. Lichtenwalner et al. (1978) reported that percentage of amylose in normal (WxWxWx), heterowaxy1 (WxWxwx i.e. single wx gene), heterowaxy2 (Wxwxwx i.e. two wx genes) and waxy (wxwxwx) sorghum were 24, 23, 17.3 and 1, respectively. Starch isolated from corneous endosperm has lower iodine binding capacity and higher gelatinization temperature than that isolated from floury endosperm; gelatinization temperature range for sorghum starch is 71 – 80 °C (Waniska and Rooney, 2000).

Protein

The second major component of sorghum and millet grains is protein. Protein content and composition varies due to genotype, water availability, temperature, soil fertility and environmental conditions during grain development. The protein content of sorghum is usually 11-13% but sometimes higher values are reported (Dendy, 1995). Grain proteins are broadly classified into four fractions according to their solubility characteristics: albumin (water soluble), globulin (soluble in dilute salt solution), prolamin or kafirin (soluble in alcohol) and glutelin (extractable in dilute alkali or acid solutions). Prolamins (kafirins) constitute the major protein fractions in sorghum, followed by glutelins. Lack of gluten is characteristic of sorghum protein composition, and makes sorghum an excellent alternative to wheat for people suffering from a wheat gluten allergy and celiac disease (US grain council).

The structural and functional properties of kafirins are reviewed by Belton et al. (2006). Belton et al 2006 have classified kafirins into four groups, called α, β, γ -kafirins (based on their relationships to the zeins revealed by their amino acid compositions and sequences, their molecular masses and their immunochemical cross-reactions), and δ-kafirin (related to the d-zeins of maize, which has been identified from the sequences of cloned DNAs but has not been characterised at the protein level).

Fat and Lipids

The crude fat content of sorghum is 3 percent, which is higher than that of wheat and rice but lower than that of maize. The germ and aleurone layers are rich in lipid fraction. The germ itself provides about 80 percent of the total fat. Neutral lipid fraction was 86.2 percent, glycolipid 3.1 percent, and phospholipid 10.7 percent in sorghum fat. Fatty acid was significantly higher in kafir, caudatum and wild sorghum than in the bicolor, durra and guinea groups. On the other hand, caudatum types had the lowest linoleic acid and bicolor, durra and guinea varieties had more than wild and kafir sorghum. Oleic and linoleic acids were negatively correlated with each other. The fatty acid composition of sorghum fat (linoleic acid 49 percent, oleic 31 percent, palmitic 14 percent, linolenic 2.7 percent, stearic 2.1 percent) was similar to that of corn fat but was more unsaturated.

6. Damage to Sorghum Grain

Around 10-30% of the production gets damaged due to unfavorable climatic conditions during growth of the sorghum plant, and inadequate transport and storage facilities. Damaged grains mean grains that are not suitable for human consumption. Damage to grains includes discoloration, breakage, cracked, damage by insect, infection by fungus, chalky appearance, partially softened by being damp, dirty and bad smell, etc. (Suresh et al. 1999, a and b). Disposal of such damaged grains is a major problem from an environmental point of view and it is also wastage of GDP. Damaged grains are not of edible quality and are available at much lower prices.

Sorghum grains suffer from infection and colonization by several mold fungi during the panicle and grain developmental stages. The infection resulting in molded grain or grain mold is normally referred to as blackening of sorghum. Many species of fungi cause grain mold in sorghum. Most grain mold fungi are relatively non-specific and can colonize several species of plants (cereals, oilseeds, spices and nuts). Aspergillus, Alternaria, Cladosporium, Diplodia, Fusarium, Curvularia, Phoma and Penicillium are among the prevalent grain mold pathogens of sorghum (Ahmed and Ravinder Reddy 1993, Bandopadyay et al. 2000, Thakur et al. 2006). Several mold-causing fungi are producers of potent mycotoxins that are harmful to health and productivity of human and animal (Bandyopadhyay et al., 2000). Hence, damage caused by insect infection and attack of fungus (blackened sorghum or grain mold) because of wet and humid weather makes sorghum grains even unfit for animal consumption.

Hence, damaged grains would be a more economical raw material for value addition through the production of bioethanol, glucose and maltose based products. Also, industrial use of healthy grains would be uneconomical because they have good market demand and obviously high cost. Industrial grade damaged sorghum grains (inclusive of 30-55% sound grains) are available in large quantity at Food Corporation of India (FCI) at 10 times lower rate than the fresh grains (Suresh et al., 1999a; 1999b).

Chandrashekar and Satyanarayana, 2006 have reviewed available information on the mechanisms of resistance to insect pests and fungal pathogens in sorghum and millets. Teetes and Pendleton (2000), Frederiksen (2000) and Stahlman and Wicks (2000) have discussed insect pests of sorghum, diseases and disease management in sorghum, and weeds and their control in grain sorghum, respectively.

7. Utilization of Sorghum

Sorghum is grown in the United States, Australia, and other developed nations essentially for animal feed. However, in Africa and Asia it is used both for human nutrition and animal feed. Utilization of sorghum can be classified mainly into the following two categories viz. 1. Human Food; and 2. Industrial use, which includes animal feed, alcohol industries, starch industry, etc. Quality of sorghum grains decides in which sector it will go. High quality sorghum grains obviously possess high cost and are used as human food. Moderate quality sorghum grains are normally used as feed for livestock or source for isolation of starch. Low quality sorghum grains which are not of edible quality or damaged grains are normally used by the alcohol industry.

7.1. Food Use for Human

Sorghum is a key staple cereal in many parts of the developing world, especially in the drier and more marginal areas of the semi-arid tropics. Per capita annual food consumption of sorghum in rural producing areas is more stable, and usually considerably higher, than in urban centers. Within these rural areas, consumption tends to be highest in the poorest, most food-insecure regions. Per capita annual consumption of sorghum and its importance as a food security crop is highest in Africa. For example, per capita annual consumption is 90–100 kg in Burkina Faso and Sudan; sorghum provides over one-third of the total calorie intake in these two countries. In Asia, sorghum continues to be a crucial food security crop in some areas (e.g., rural Maharashtra in India, where per capita annual consumption is over 70 kg) (ICRISAT/FAO, 1996).

In 1979-81, 39% of global production (65.4 million tonnes) was used as human food and 54% for animal feed, whereas in 1992–94, 42% of total utilization (63.5 million metric tonnes) was for human food and 48% for animal feed. The proportion of food utilization has gradually increased as a result of a greater food use in Africa and the substitution of sorghum by other grains (mainly maize) as feed elsewhere (Source: ICRISAT/FAO, 1996).

However, both production and food utilization have fallen during the 1980s and early 1990s, because of shifting consumer preferences. As incomes rise, consumers are shifting to wheat and rice which taste better and are easier and faster to cook. This trend is accentuated by rapid urbanization and the growing availability of a range of convenience foods based on wheat and rice. ICRISAT/FAO (1996) have discussed sorghum economy in detail. Several previous reviews have addressed the subject of traditional foods from sorghum in depth, for example Murty and Kumar (1995), Rooney and Waniska (2000), and Rooney and Serna-Saldivar (2000).

There is a wide array of traditional sorghum products that currently exist in areas where sorghum production is abundant, namely Africa and India. While there are many regional variations, the categories of these products remain fairly consistent. Table 4 shows groupings of these traditional products compiled by Waniska and Rooney (2000), and modified by Schober et al. (2006).

Sorghum is eaten in a variety of forms that vary from region to region. In general, it is consumed as whole grain or processed into flour, from which traditional meals are prepared. There are four main sorghum-based foods:

Table 4. Traditional sorghum products

Food	Example	Region of Origin
Unfermented Bread	Roti, chapati	India, East Africa
Fermented Bread	Dosa	Southern India
Stiff Porridge	Ugali	Southern/ East Africa
Thin Porridge	Uji	East Africa
Steamed Products	Couscous	West Africa
Alcoholic Beverages	Dolo	West Africa
Sour/Opaque Beer	East Africa	East Africa

Adapted from Waniska and Rooney (2000), Schober et al. (2006).

- Flat bread, mostly unleavened and prepared from fermented or unfermented dough in Asia and parts of Africa;
- Thin or thick fermented or unfermented porridge, mainly consumed in Africa;
- Boiled products similar to those prepared from maize grits or rice;
- Preparations deep-fried in oil.

The most common and simplest food prepared from sorghum and millets is porridge. In all cultures traditionally depending on cereals, a range of treatments of the whole seed before milling and sifting has been applied. The treatment procedures are steeping, fermentation, malting, alkali or acid treatment, popping, roasting (dry or wet), parboiling, and drying. One of the aims of seed treatment is to remove the polyphenolic compounds from the seed. Others are to improve storage quality, or to make many kinds of snacks and other popular foods. The traditional art of food preparation is not standardized and routine procedures have been passed on to the women through generations.

The main foods preparations with sorghum are: tortillas (Latino America), thin porridge, e.g. "bouillie" (Africa and Asia), stiff porridge, e.g. tô (West Africa), couscous (Africa), injera (Ethiopia), nasha and kisra (Sudan), traditional beers, e.g. dolo, tchapallo, pito, burukutu, etc. (Africa), ogi (Nigeria), baked products (U.S.A., Japan, Africa), etc. Tortillas are a kind of chips prepared from sorghum alone or by mixing sorghum with maize and cassava (Anglani, 1998). Nasha is a traditional weaning food (infant porridge) prepared by fermentation of sorghum flour. Ogi is an example of traditional fermented sorghum food used as a weaning food, which has been upgraded to a semi-industrial scale. Injera is a local fermented pancake-like bread prepared in Ethiopia from sorghum. Kisra is traditional bread prepared from the fermented dough of sorghum (Reviewed by Dicko et al. 2006).

Often sorghum porridges are characterized by thick pastes that may form rather stiff gels depending on the variety used. Porridges prepared with malted sorghums have several order of magnitudes lower viscosities than those of non-malted sorghums. These porridges are particularly useful for the formulation of weaning foods for infants because of their high energy density. Furthermore the use of exogenous sources of a-amylase from higher West African plants is useful in reducing the viscosity of cereal porridges, including sorghum ones.

The stiff porridge prepared from maize or cereal mixture (maize, sorghum, pearl millet, finger millet, etc.) in Kenya, Uganda and Tanzania is commonly called *ugali*. In much of Northern Africa a steamed, granulated product called couscous, made from cereal flours (mostly wheat) is highly popular. In West Africa, sorghum, pearl millet, maize, and fonio are used to prepare couscous, although pearl millet is preferred. Sorghum noodles are an important food product in China. Sorghum is used for tortilla preparation either alone or in combination with maize in Honduras, Nicaragua, Guatemala, El Salvador and Mexico. Roti or bhakri is an unfermented dry roasted pancake made in India from wheat, sorghum, pearl millet and maize flour. In addition, pasta products, such as spaghetti and macaroni made from semolina or wheat could be made with mixtures of composite flour consisting of 30-50% sorghum in wheat (Hugo et al., 2000, 2003).

Whole or decorticated sorghum is often consumed as a rice-like product in several African countries: in Sudan, "pearl dura"; in Kenya, "supa mtama"; and South Africa, "corn rice" (Taylor and Dewar 2001). The main drawback to these products is their long cooking time. For example, South African corn rice must be rinsed, pre-cooked for 10 to 15 minutes, rinsed again, and then cooked for another 25-35 minutes (Perten 1983). Another rice-type product made with sorghum is couscous. While couscous is traditionally known in the U.S. as a short grain pasta product made from wheat semolina, sorghum couscous is a steamed, agglomerated food made from the flour of decorticated sorghum (Anglani 1998). Like porridges, couscous is served for breakfast, lunch or dinner with either buttermilk or more savory items such as fish or vegetables.

Sorghum grain is used in the production of two types of beer: clear beer and opaque beer. The latter is a traditional, low-alcohol African beer that contains fine suspended particles. Dolo is a reddish, cloudy or opaque local beer prepared essentially from red sorghum malt (Hilhorst, 1986; as cited in Dicko et al. 2006). The primary quality criterion of selection of sorghum varieties for beer is their potential to produce malt with high α-amylase and β-amylase activities (Verbruggen, 1996; Taylor and Dewar, 2001). Sorghum is traditionally a major ingredient in home-brewed beer. Small quantities are used in the beer industries in Mexico and U.S.A.. Sorghum is a good source of starch, cellulose, and glucose syrup. Although domestication was primarily for food (and also for beer and sweet stems in Africa, and for brooms in China), crop residues have been valued as animal fodder, building materials, and fuel. By applying hydrothermic technologies (flaking, puffing, extrusion, micronizing) new sorghum and millet products of good quality and good taste can be produced (Leder, 2004). These beverages are characterized by a sour, lactic acid flavor, provided either by lactic acid fermentation or by the addition of commercially produced lactic acid (Taylor and Dewar 2001).

Dicko et al. (2006) has reviewed and emphasized on the impact of starch and starch depolymerizing enzymes in the use of sorghum for some African foods. Dicko also discussed the feature of sorghum as a food in developing as well as in developed countries. Taylor et al. (2006) have reviewed the role of sorghum in nutrition and health of humans, and literature on novel and non-traditional sorghum foods like Gluten-free leavened breads (starch bread and additives, flour breads and additives, effect of cultivar and theoretical basis for sorghum functionality in gluten-free bread making), cakes and cookies, Tortillas, snack foods, parboiled sorghum, and noodles. In the U.S.A. and Japan, sorghum utilization as human food is increasing because of its use in snacks and cookies (Rooney and Waniska, 2000).

In the developed countries, nowadays there is a growing demand for gluten-free foods and beverages from the people suffering from celiac disease and other intolerances to wheat, who cannot eat products from wheat, barley, or rye. Celiac disease is a syndrome characterized by damage to the mucosa of the small intestine caused by ingestion of certain wheat proteins and related proteins in rye and barley. The gliadins and glutenins of wheat gluten have been shown to contain protein sequences that are not tolerated by celiac. The average worldwide prevalence has been estimated as high as 1:266. Estimates place the number of persons with celiac disease in the U.S.A. at roughly 3 million. The cornerstone treatment for celiac disease is the total lifelong avoidance of gluten ingestion. This means that wheat, rye, and barley have to be avoided, including durum wheat, spelt wheat, kamut, einkorn, and triticale. Sorghum is often recommended as a safe food for coeliac patients, because it is only distantly related to the Triticeae tribe cereals wheat, rye and barley, being a member of the Panicoideae sub-family which also includes maize and most millets. Sorghum therefore, provides a good basis for gluten-free breads and other baked products like cakes and cookies (biscuits) and in snacks and pasta (Taylor et al., 2006). Sorghum alone is not considered as a bread making cereal because of the lack of gluten, but addition of 20-50%

sorghum flour to wheat flour produces excellent bread (Anglani, 1998; Carson et al., 2000; Hugo et al., 2000, 2003). Very little research (Scober et al. 2005, Scober et al. 2007) has been done on gluten-free sorghum breads. Such breads would be suitable for those on a gluten-free diet, as well as providing cost savings and a possible replacement for wheat breads in developing countries where wheat is not the predominately native cereal. Recently, Fredrick (2009) has studied the effect of sorghum flour composition and particle size on quality of gluten-free bread.

Sorghum can contain substantial levels of a wide range of phenolic compounds, which have health promoting properties, in particular their antioxidant activity. Their use as nutraceuticals and in functional foods are reviewed in the paper by Dykes and Rooney (2006). In addition to the potential health benefits of sorghum phenolics, sorghum wax may also have unique health properties. Long-chain fatty alcohols, aldehydes and acids are interconverted in cellular metabolism, so that all three classes might lower cholesterol. Policosanols (fatty alcohols in sorghum wax) are a promising resource for the prevention and therapy of cardiovascular disease. Crude lipid extract from whole kernel sorghum, which comprised a wide range of lipid substances including plant sterols and policosanols, lowered cholesterol absorption and plasma non-HDL cholesterol in hamsters. (Taylor et al., 2006).

7.2. Industrial Utilization

The main industries using sorghum are the animal feed sector, alcohol distilleries, and starch industries.

7.2.1. Animal Feed

About 48 percent of world sorghum grain production was fed to livestock (human food use constitutes about 42 percent). In contrast to food utilization, which is relatively stable, utilization of sorghum as animal feed changes significantly in response to two factors: rising incomes, which stimulate the consumption of livestock products, and the price competitiveness of sorghum vis-à-vis other cereals, especially maize. While sorghum is generally regarded as an inferior cereal when consumed as food, the income elasticities for livestock products (and hence the derived demand for feed) are generally positive and high.

Demand for animal feed is concentrated in the developed countries and in middle-income countries in Latin America and Asia, where demand for meat is high and the livestock industry is correspondingly intensive. Over 85 percent of the use of sorghum as animal feed occurs in Developed countries (ICRISAT/ FAO).

Another important factor is consumer preference for meat color. Maize contains higher carotene levels than sorghum, so meat from maize-fed animals tends to be more yellow than meat from sorghum fed animals. In Japan for example, consumers generally prefer white-coloured meat. Therefore, sorghum is a valued ingredient in some compound feed rations (for poultry, pigs and some breeds of beef cattle). In contrast, sorghum is discounted by producers in India because consumers there generally prefer poultry meat and egg yolks with a deeper yellow colour (ICRISAT/ FAO).

Table 5. Industry-perceived advantages and disadvantages of using sorghum in poultry feed (Kleih et al., 2007)

Advantages	Disadvantages
• Low cost	• Lower energy content than maize
• Energy alternative to maize	• Risk of aflatoxins (often associated with blackened grain)
• Easy availability	• Risk of tannins
• Good pelletability	• Not always available
	• Problems with grinding; mash becomes powdery reducing feed intake by birds
	• Low palatability and digestibility
	• Varying quality; grain often infested with weevils, fungi, etc.
	• Sorghum lacks the carotenoid pigments present in yellow maize, which are necessary for egg yolk colour
	• Feed including sorghum is more difficult to
	• Absence of standard varieties in the market

Sorghum grain is used in poultry feed in small quantities, although maize is the major cereal for poultry feed. Sorghum can replace maize from 50% to 74% (Thakur et al. 1984; Rama Rao et al. 1995). Sorghum grain can replace maize in poultry feed to a great extent in view of the similarity in chemical composition of the grain. The results on egg production and broiler weight were similar in the two experiments, when sorghum or maize was fed as a source of energy. Local sorghum grain was as effective as high-yielding sorghum or maize. Thus, sorghum grain has high potential for use in poultry feed (Subramanian and Metta, 2000).

Table 6. Advantages and disadvantages of using sorghum in alcohol production (Kleih et al., 2007)

Advantages	Disadvantages
No major technical constraints with modern technology	Sorghum is a food grain, and may not be available for alcohol production in times of food shortages
Causes least pollution	Some producers in Maharashtra face difficulties in selling grain-based alcohol, largely due to the State-imposed export pass fee. This difficulty is localized
Good quality alcohol free from sulphates and aldehydes present in molasses based alcohol	Cost of molasses based alcohol is lower than grain based alcohol
Can create demand for damaged grain	
Possible regular sourcing of grain from rainy-season crop	
Byproduct of grain alcohol production can be used as animal feed	

The limited inclusion of sorghum in poultry feed and its relatively low status as a raw material can be accounted to disadvantages of sorghum as given in the Table 5.

7.2.2. Alcohol Industries

Sorghum has the potential for being used in the production of bio-industrial products, including bioethanol. Sorghum is a starch-rich grain with similar composition to maize, and, as with all cereals, its composition varies significantly due to genetics and environment. Starch ranges of 60–77% and 64–78% have been reported for sorghum and maize, respectively. As such, sorghum grain would be appropriate for use in fermentation similar to the use of maize for the production of bioethanol. Its use may be of particular benefit in countries where rainfall is limiting and maize does not grow well. Taylor et al. (2006) concluded from the available data that 1.2–2.3 million metric tonnes of sorghum was used for ethanol production, 3.7–7.5% of the grain used for ethanol production was sorghum, and 0.13–0.25 billion gallons (0.49–0.95 billion litres) of ethanol originated from sorghum.

While discussing the potential for using sorghum in alcohol production, one must keep in mind that in India, molasses (a byproduct of sugar manufacture using sugarcane) constitutes the most important raw material in this industry. It is estimated that about 95% of the alcohol manufactured in India is from molasses and the rest comes from grains, roots and tubers. Although, the quantity of sorghum grain presently used by the alcohol sector in India is comparatively low (Table 6), it seems to be the most "enthusiastic" user of the crop as an industrial raw material. Nowadays government policies on licensing alcohol production and trade are changing and also government is promoting production of grain based alcohol. Hence, the present scenario is providing an opportunity for sorghum to gain greater acceptability as a raw material in the alcohol industry. Some distillers indicated a preference for varieties with a higher starch content and less protein. Distilleries had no objection to using severely blackened grain as long as the starch content was acceptable. In general, like most other industrial users, distilleries purchase rainy-season sorghum through traders or brokers in main producing centers. The problem with this system could be the misuse of the position by brokers to "control" the market. In this context, contract farming may be an option providing better linkages between producers and industrial users (Kleih et al., 2007).

Maharashtra Government is in favor of promoting grain-based alcohol production to create a demand for rainy-season sorghum. It must be remembered in this context that rain-damaged or blackened sorghum could be a favorable raw material for alcohol production because of its lower market price. Maharashtra, the main producer of rainy season sorghum, regularly faces the problem of finding suitable users of blackened sorghum which constitutes 40-60% of its produce, depending on the rainfall pattern during grain maturity. Advantages and disadvantages of using sorghum in alcohol production are given in the Table 6. Though they are given from the perspective of its use in India, especially Maharashtra, they are universal in nature.

Demand of ethanol is consistently increasing as an alternative energy source. Ethanol is normally manufactured from sugarcane molasses. This conventional substrate is no longer a cheap and good quality raw material due to its decontrol by Indian government. Also in order to compete in the international market, there is the need to improve the alcohol quality. Hence, since the 1990s industries are diverting from use of molasses to starchy substrates for production of ethanol. Also, due to several environmental issues associated with production of ethanol from molasses, government is promoting grain based alcohol. In this section, literature review is only limited to use of sorghum for ethanol production and not the use other starchy substrates like corn, wheat, tapioca, rice, etc.

In typical grain fermentation the alcohol yield is >360 L/t of grain, containing 60% starch which corresponds to 85% of conversion of sugars to alcohol. During fermentation grain starch gets utilized and the other components such as cellulose and protein remain unutilized in the stillage. Moreover, the remaining sugars/starch do not go into the effluent because the stillage is dried. The dried residue is called Distiller's Dried Grain and Solubles (DDGS), which is an excellent ingredient for animal feed. The dried stillage, DDGS, is not only rich in protein and fat, but also rich in vitamins produced by yeast during fermentation. Approximately 240–260 kg DDGS/t of grain is produced as a byproduct. DDGS makes an excellent animal feed. Therefore, all residues are utilized. Utilization of damaged grains for ethanol production offer value addition to damaged grains and can create a market for damaged grains thereby considerably helping farmers producing sorghum (Shaorin et al. 2000).

Literature Review on Use of Sorghum as Source for Ethanol

There is a good quantity of literature available on the utilization of sorghum as a source to produce ethanol. Application of research findings to bioprocessing of sorghum grain could benefit both grain producers and the bio-industry via the following areas: (1) approaches and capabilities to further improve the efficiency of sorghum processing; (2) improvement in sorghum conversion yield to industrial products, thereby improving sorghum economics; (3) information to assist in the development of new and improved sorghum hybrids; and (4) enhancement of economic rural development through expanded sorghum production, especially across the many drier sorghum-growing states (Wang et al 2008). According to Wang et al. 2008 grain sorghum is a viable feedstock for ethanol production. They also confirmed that ethanol yield increased with starch content, however, low amylose grains are preferred for ethanol production via simultaneous saccharification and fermentation (SSF) because starch in the amylose–lipid complex cannot be converted into fermentable sugar; sorghums with the highest degree of protein cross-linking had the lowest fermentation efficiency; high-tannin varieties are not a good choice for ethanol production. They have developed an energy life cycle analysis model (ELCAM) to quantify and prioritize the saving potential from factors identified in this research and used it to identify factors that most impact sorghum use.

Suresh et al. (1999a) developed a simultaneous saccharification and fermentation (SSF) system for producing ethanol from damaged sorghum (50% sound and 50% damage grains). They have reported ethanol yields of 91.5% and 78.6% of the theoretical ethanol yield with use of VSJ1 strain and standard strain MTCC 170 for damaged sorghum. These authors later utilized a similar SSF method to compare ethanol production from damaged (50% sound and 50% damaged grains) and high quality sorghum (Suresh et al., 1999b). It must be noted that the latter method involved no cooking step. Raw flour starch was saccharified by Bacillus subtilis amylase and fermented by Saccharomyces cerevisiae. The damaged portion included kernels that were broken, cracked, attacked by insects, dirty or discolored. The high-quality sorghum flour was obtained locally. They found that using a level of 25% (w/v) substrate yielded 3.5% (v/v) ethanol from the damaged grain sample. For comparison, the high-quality sorghum flour yielded 5.0% (v/v) ethanol. The values of optimum pH and temperature were reported to be 5.8 and 35 °C respectively for SSF process for damaged sorghum. The damaged grain sample was reported to be ten times cheaper than high-quality grain and thus may be an economical way to produce ethanol even though yields were lower. The authors further emphasized that utilization of raw starch (i.e. without cooking) would save energy.

Zhan et al. (2003) investigated the impact of genotype and growth environment on the fermentation quality of sorghum. Eight sorghum hybrids grown in two different locations were used to produce ethanol. The process included the following steps: heating with thermostable α-amylase at 95 °C and then 80 °C (liquefaction), incubation with amyloglucosidase at 60 °C (saccharification), inoculation with S. cerevisiae and fermentation for 72 h at 30 °C. It was found that ethanol concentrations varied relatively narrowly (about 5%) across the 16 samples. Genotype and production environment had a significant effect on chemical composition and physical properties of the sorghum used in this study, which in turn significantly affected ethanol yields. The correlation between ethanol concentration and starch content was positive, as expected, but low ($r = 0.35$, $P > 0.05$), while a much more distinct negative correlation between ethanol concentration and protein content was found ($r = -0.84$, $P < 0.001$). Since protein and starch content are inversely proportional, it is not surprising that opposite correlations for these two measures to ethanol production would be found. But, protein content does not have a significant effect on the percentage of the theoretical ethanol yield. However, it is interesting that protein had a much stronger relationship to ethanol yield than did starch. More research is needed to determine exactly what components in the grain, and their interactions, are responsible for ethanol yields in sorghum. It is possible that during the initial heating steps, a disulphide-mediated protein polymerization process occurred, resulting in web-like or sheet-like protein structures, as described by Hamaker and Bugusu (2003). Under these conditions, some of the starch might be trapped in these protein webs, and its full gelatinisation and degradation by amylases might be hampered. Evidence for this is provided by the work of Zhan et al. (2006) who investigated cooking sorghum using supercritical-fluid-extrusion (SCFX) to gelatinize the starch. In SCFX supercritical carbon dioxide is used in place of water as the blowing agent. Using SCFX increased ethanol yields by around 5% compared to non-extrusion cooked sorghum. An improvement in the bioconversion of sorghum starch was accounted to the release of starch from the protein matrix due to SCFX and enhancing the availability of starch for conversion to fermentable sugar. Literature review on ethanol production is summarized in Table 7 with reaction conditions and remarks as parameters. Recently, Mojovic et al. 2009 has reviewed progress in production of bioethanol on starch based feedstocks and Taylor et al. (2006) has reviewed literature on production of ethanol from sorghum.

In addition to breeding sorghum specifically for fermentation quality, pre-processing the grain can be used to improve ethanol yields and process efficiency. Corredor et al. (2006) investigated decorticating sorghum prior to starch hydrolysis and ethanol fermentation. In general, decortication decreased the protein content of the samples up to 12% and increased starch content by 5–16%. Fiber content was decreased by 49–89%. These changes allowed for a higher starch loading for ethanol fermentation and resulted in increased ethanol production. Ethanol yields increased 3–11% for 10% decorticated sorghum and 8–18% for 20% decorticated sorghum. Using decorticated grain also increased the protein content of the distillers dried grains with solubles (DDGS) by 11–39% and lowered their fiber content accordingly. Using decorticated sorghum may be beneficial for ethanol plants as ethanol yield increases and animal feed quality of the DDGS is improved. The bran removed before fermentation could be used as a source of phytochemicals (Awika et al., 2005) or as a source of kafirin and wax.

Table 7. Summary of literature review on ethanol production from sorghum

Reference	Reaction conditions	Remarks
Suresh et al., 1999a	200 mL slurry autoclaved at 121 °C for 30 min and a 3% inoculum of A. niger (NCIM 1248) and 7% inoculum of yeast was added to it. 2-12% starch concn, 150 rpm, 30 °C, 5 days	Damaged sorghum (50% sound and 50% damaged grains used in this work)
Suresh et al., 1999b	100 ml slurry and 0.3% peptone 0.1% KH_2PO_4 and 0.1% $(NH_4)_2SO_4$; pH 5.8. The crude amylase broth (10 ml) of B. subtilis VB2 and 6% S. cerevisiae VSJ4 suspension was added to slurry and incubated at 35 °C and 200 rpm for 4 days.	Optimized conditions being pH 5.8, 35 °C and 25% w/v slurry concn. High quality and damaged sorghum produced 5% v/v and 3.5% v/v ethanol production with no cooking step.
Zhan et al., 2003	100 mL slurry, pH 5.8. Liquefaction: 95 °C for 45 min, 80 °C for 30 min (0.01 mL amylase/g of starch in both steps of liquefaction). Saccharification: 60 °C (150 U/g of starch) for 30 min. 50 rpm for all steps. Fermentation: 20 g ground sorghum, 0.3 g peptone, 0.1 g KH_2PO_4, and 0.1 g $(NH_4)_2SO_4$ at pH 3.8. Medium was inoculated with 6% yeast suspension (1×10^6 cells/mL) and incubated 200 rpm for 72 h at 30 °C.	16 different varieties of sorghum. Positive correlation between ethanol concn and starch content and negative correlation between ethanol concn and protein content.
Zhan et al., 2006		Extrusion could break disulphide protein bonds and disrupt the protein matrix, gelatinize starch, and make more starch available for enzyme hydrolysis, and consequently, increase ethanol yield and fermentation efficiency.
Wu et al., 2007	30 g flour was mixed with 100 mL distilled water. 10 µL liquozyme was added and slurry was digested for 45 min at 95 °C. Slurry cooled to 80 °C and second dosage of 10 µL liquozyme was added and liquefaction was continued for additional 30 min at 120 rpm. Saccharification: 100 µL Spirizyme, 120 rpm, 60 °C, 30 min. pH adjusted to 4.2 and inoculated with 5 mL of 48 h yeast pre-culture.	Ethanol yields varied by 22% and conversion efficiencies by 9% among 70 sorghum samples. Positive effect of starch content on fermentation efficiency and negative effect of protein, tannin, crude fiber, and ash content on fermentation efficiency was observed.
Zhao et al., 2008		During mashing cross-linked microstructure, which could hold starch granules or polysaccharides inside or retard or prevent the access of enzymes to starch get formed. Severe cross-linking in mashed sample was most likely because of a combination of heat-induced cross linking and cross-linking because of protein-tannin interactions. Tight and open microstructures were observed with low conversion sorghum and high conversion sorghum, respectively.

Wu et al. (2007) have performed ethanol production from 70 genotypes and elite hybrids of sorghum using dry grind process and identified factors impacting ethanol production. They have observed variation in the ethanol yield by 22% and conversion efficiencies by 9% among 70 sorghum samples, indicating significant effect of sorghum genotype on fermentation efficiency to ethanol. They reported a positive effect of starch content on fermentation efficiency and negative effect of protein, tannin, crude fiber, and ash content on fermentation efficiency. Protein digestibility of waxy, normal sorghum (60-68 %) were higher that that for high tannin sorghum (28%); higher fermentation efficiencies were observed for waxy, normal sorghum (89-90%) than those of high tannin sorghum (85%). After mashing sorghum protein appeared to produce highly extended, strong web like micro structures (in accordance with results of Hamaker et al., 2003) into which small starch granules were tightly trapped. These changes related to protein structure during mashing could contribute to incomplete gelatinization and hence hydrolysis of starch and conversion efficiency to ethanol. DSC thermograms of waxy sorghum starch consists of single endothermic peak (60-80 °C), whereas that of normal sorghum starch showed the presence of two endothermic peaks; one, in 60-80 °C corresponds to amylopectin, and the second, in 85-105 °C corresponds to amylose-lipid complex. Waxy sorghum gives higher conversion efficiencies to ethanol than normal sorghum; this mainly happens due to the presence of amylose-lipid complex. Wu et al. (2007) concluded that the major factors adversely affecting conversion efficiency to ethanol being condensed tannin, high viscosity, low protein digestibility, protein-starch interactions, and amount of amylose-lipid complex.

Zhao et al. (2008) have characterized the changes in sorghum protein in digestibility, solubility, and microstructure during mashing and to relate those changes to ethanol fermentation quality of sorghum by using 9 sorghum cultivars (2 contained tannin and the rest were tannin free). They have observed that protein solubility and *in vitro* protein digestibility decreased significantly after mashing. Tendency of sorghum proteins to form highly extended, strong web like microstructures during mashing was confirmed using CFLSM (confocal laser-scanning microscopy) images. Formation of web like structure was earlier reported by Hamaker and Bugusu (2003) and Wu et al. (2007). They reported that the cultivar with the lowest conversion efficiency formed a tightly cross-linked microstructure, which could hold starch granules or polysaccharides inside or retard or prevent the access of enzymes to starch, and severe cross-linking in this sample was most likely because of a combination of heat-induced cross linking and cross-linking because of protein-tannin interactions. More open web-like microstructures were observed in cultivars with higher conversion efficiencies upon mashing. Protein digestibility of the unmashed sorghum, solubility and the SE-HPLC area of proteins extracted from mashed samples were highly correlated with ethanol fermentation. Since protein cross linking plays a significant role in the fermentation, it was expected that γ-kafirin (%) would relate significantly with conversion efficiency. But, it neither correlated to ethanol yield nor conversion efficiency significantly. They concluded that protein cross-linking does play a role in the production of ethanol from sorghum, albeit through indirect measures of protein cross-linking (i.e. reduction in protein digestibility after mashing, which is due to protein cross linking).

7.2.3. Starch Industry

In general the wet milling of sorghum is similar to that of maize. A thorough review of early research on wet milling of sorghum can be found in Munck (1995). However, a problem particular to sorghum is the presence of polyphenolic pigments, the pericarp and/or glumes, which stains the isolated starch (Beta et al., 2000a–c). Due to this reason sorghum is not very popular in starch producing industries. Sorghum is used for starch production only when maize is not available. Recently, Taylor et al. (2006) have shortly reviewed literature on starch production from sorghum. Sweetener and fermentation industries are two of the main consumers of the starch. Nutritive sweeteners are mainly products of enzymatic hydrolysis of starch, namely maltodextrins, high maltose syrup, maltose, glucose syrup, dextrose, and fructose, which are used in the food and pharmaceutical industries. Isolated starch from sorghum has the same applications as corn starch or any other starch. There are two excellent textbooks (Schenck and Hebeda, 1992; Kearsley and Dziedzic, 1995) on the basic concepts of the production of starch hydrolysate. Three steps in which enzymes are used in starch hydrolysis and processing are as follows: 1. Liquefaction of starch (combination of starch gelatinization and dextrinization of gelatinized starch) 2. Saccharification of starch liquefact (containing oligosaccharides) to glucose or maltose 3. Isomerization of glucose. Different types of sweetener that can be produced from isolated starch are given in Table 8.

In 2004, starch production in the world was around 60 million t. 70 % of the starch was produced from corn, other raw materials being wheat, sweat potato, cassava and potato (www.starch.dk/isi/stat/rawmaterial.html). In India, 0.7 million t of starch was produced in 1998, out of which 0.6 million t was produced from maize and 0.1 million t was from cassava (Kleih et al., 2007).

In recent years several new developments in sorghum wet milling have been reported. Perez-Sora and Lares-Amaiz (2004) investigated alkaline reagents for bleaching the starch and found a mixture of sodium hypochlorite and potassium hydroxide to be the most effective. To improve the economics of sorghum wet-milling Yang and Seib (1995) developed an abbreviated wet-milling process for sorghum that required only 1.2 parts fresh water per part of grain and that produced no waste water. The products of this abbreviated process were isolated starch and a high moisture fraction that was diverted to animal feed.

Buffo et al. (1997) investigated the impact of sulphur dioxide and lactic acid steeping on the wet-milling properties of sorghum and reported that the amount of lactic acid used during steeping had the most impacted wet-milling quality characteristics such as starch yield and recovery. These authors also investigated the relationships between sorghum grain quality characteristics and wet-milling performance in 24 commercial sorghum hybrids (Buffo et al., 1998). Perhaps not surprisingly, they found that grain factors related to the endosperm protein matrix and its breakdown and subsequent release of starch granules as important factors in wet-milling of sorghum. Related to this, Mezo-Villanueva and Serna-Saldivar (2004) found that treatment of steeped sorghum and maize with protease increased starch yield, with the effect being greater on sorghum than maize.

Wang et al. (2000) optimized the steeping process for wet-milling sorghum and reported the optimum steeping process to utilize 0.2% sulphur dioxide, 0.5% lactic acid at a temperature of 50 °C for 36 h. Using this steeping process, wet-milling of sorghum produced starch with a lightness value (Black and white samples will have lightness value of 0 and 100 respectively. Lightness value is measured by chroma meter and normally denoted as L) of 92.7, starch yield of 60.2% (db), and protein in starch of only 0.49% (db). Beta et al. (2000a–c) found that both polyphenol content and sorghum grain properties influence sorghum starch properties.

Table 8. Different types of nutritive sweetener

Nutritive sweetener	Dextrose equivalent (DE)	Class (Degree of polymerization)
Maltodextrins	<20	Malto-oligosaccharides (3-9)
High maltose syrup	20-50	Maltose (2), glucose (1), Malto-oligosaccharides
Maltose	53	Sugar (2)
Glucose syrup	20-95	Glucose, maltose, malto-oligosaccharides
High fructose syrup	-	Fructose, glucose (1)
Dextrose	100	Sugar (1)

Using sorghum grits as the starting material for wet-milling rather than whole sorghum produced lower yields, but the isolated starch was higher in quality. Sorghum starch matching the quality of a commercial corn starch was successfully produced by wet-milling sorghum grits (Higiro et al., 2003).

Xie and Seib (2002) developed a limited wet-milling procedure for sorghum that involved grinding sorghum within the presence of 0.3% sodium bisulphite solution. This procedure produced starch with an L value of 93.7 and a starch recovery of 78%. Large grain sorghum hybrids wet-milled by this "no steep" procedure were reported to produce high-quality starches with L values from 93.1 to 93.7 (compared to 95.2 for a commercial corn starch). Some of the large grain hybrids tested showed promise for easy recovery of the germ by flotation in a similar fashion as is done for maize (Xie et al., 2006).

Park et al. (2006) reported the use of ultrasound to rapidly purify starch from sorghum. This procedure resulted in very high-purity starch with only 0.06% residual protein in the starch. New developments in wet-milling procedures for sorghum as well as breeding sorghum hybrids with improved wet milling characteristics should be of benefit for the industrial use of sorghum starch, either directly for the production of bioethanol or other industrial uses such as the production of activated carbon (Diao et al., 2002) or isolation of phytosterols from wet-milled fractions (Singh et al., 2003).

7.2.4. Other Industrial Applications

Malting and Brewing

Malting and brewing with sorghum to produce lager and stout, often referred to as clear beer as opposed to traditional African opaque beer, has been conducted on a large, commercial scale since the late 1980s, notably in Nigeria. Nigeria brews in excess of 900 million litres of beer annually. Brewing with sorghum is now also taking place in east Africa, southern Africa, and the U.S.A. (Taylor et al., 2006).

In India, easily available barley malt is preferred as the principal raw material for brewing. Sorghum is not currently used for beer production in India either as malt or as an adjunct (Kleih et al., 2007).

There has been extensive research and development work and several excellent reviews published covering enzymes in sorghum malting, sorghum malting, and brewing technology (Agu and Palmer, 1998; Hallgren, 1995; Owuama, 1997, 1999; Taylor and Dewar, 2000, 2001). Major outstanding problem areas (like use of tannin sorghum in malting and brewing, starch gelatinization, the role of the endosperm cell walls and beta-amylase activity in malt) in sorghum brewing that are specific to characteristics of sorghum grain and sorghum malt are reviewed by Taylor et al. (2006).

Film Formation

Nowadays, the environmental concerns are growing; therefore, the researches to develop edible and biodegradable films from natural renewable sources have accelerated. Edible films with the quality of renewability, degradability, compostability, and edibility could make environment friendly films particularly appealing for food and nonfood packaging applications (Musigakun and Thongngam, 2007).

Nowadays, zein is used as edible film for several products such as coating on fruits or meat to keep them fresh and reduce water loss (Trezza and Vergano, 1994). Buffo et al. 1997 used laboratory-extracted sorghum kafirin to produce films, which had tensile and water vapor barrier properties similar to films from commercial corn zein. Buffo anticipated that advancements in the kafirin extraction process will improve the film-forming ability and film properties, including color characteristics, of kafirin and indicated that sorghum kafirin has the potential to be used as a biopolymer for edible or nonedible film and coating applications.

The major protein in sorghum is prolamin also called kafirin (Chandrashekar and Mazhar, 1995). Kafirin is similar to zein in its molecular weight, solubility, structure and amino acid composition (Da Silva and Taylor, 2005). Musigakun and Thongngam, 2007 have studied characteristics and functional properties of sorghum protein (kafirin) from different sorghum cultivars to understand and be able to use kafirin as an edible film. Sorghum kafirin is an alternative source for zein in the preparation of edible film (Musigakun and Thongngam, 2007). Kafirin is more hydrophobic and has more disulfide bonds than zein; therefore it could make a better barrier and stronger films (Taylor et al., 2005).

Emmambux et al. 2004 has used tannic acid as a modifier to change kafirin film properties. Taylor and Taylor (2009) have concluded that cost effectiveness of sorghum lager beer brewing can be improved by adding value to a co-product like brewers spent grain through extraction of kafirin, which can be further used for bioplastic and microparticle production. They have used acetic acid for extraction of proteins from whole sorghum grain flour and sorghum brewers spent grain and reported that isolated protein purer kafirin because water soluble proteins would be removed by brewing process.

7.3. Value Addition to Damaged Sorghum

Though production of sorghum is high, demand for sorghum is decreasing with change in the way of living due to increased urbanization, increased per capita income of the population, and easy availability of other preferred cereals in sufficient quantities at affordable prices. Hence, in addition to being a major source of staple food for humans, it also serves as a source of feed for cattle and other livestock in scarcity of maize, but at lower prices. Also, about 10-20% of the production gets wasted due to damage and inadequate transport and storage facilities. Industrial grade damaged sorghum grains (inclusive of 30-55% sound grains) are available in large quantity at Food Corporation of India (FCI) at 10 times lower rate than the fresh grains (Suresh et al., 1999a). Damage includes chalky appearance, cracked, broken, mold, infection, etc. These damaged grains are not suitable for human consumption. Several mold-causing fungi are producers of potent mycotoxins that are harmful to the health and productivity of humans and animals (Bandyopadhyay et al., 2000). Hence, damage caused by insect infection and attack of fungus (blackened sorghum or grain mold) because of wet and humid weather makes sorghum grains even unfit for animal consumption.

Hence, an industrial application is needed to be exploited for normal and blackened sorghums in order to make sorghum cultivation economically viable for farmers, through value added products. There is a very small amount of research done on value addition to sorghum through production of glucose (Devarajan and Pandit, 1996; Aggarwal et al., 2001; Shewale and Pandit, 2009), production of ethanol (Wu et al., 2007; Suresh et al., 1999a, 1999b; Zhan et al., 2003; Zhan et. al., 2006; Zhao et al., 2008) and isolation of starch (Yang and Sieb, 1996; Xie and Seib, 2002; Higiro et al., 2003; Perez-Sira and Amaiz, 2004; Park et al., 2006). The reason for the lower level of industrial exploitation can be attributed to reduced sorghum starch digestibility (Lichtenwalner et al., 1978; Rooney and Pflugfelder, 1986; Chandrashekar and Kirleis, 1988; Zhang and Hamaker, 1998; Elkhalifa et al., 1999; Ezeogu et al., 2005) and reduced protein digestibility (Duodu et al., 2003) after cooking i.e. heat-moisture treatment of sorghum flour. Literature related to production of ethanol, isolation of starch from sorghum, and digestibility of sorghum starch and sorghum proteins is reviewed in sections 7.2.3, 7.2.4, 8, respectively. Siruguri et al. 2009 recently showed that fermented products that have nutritional benefit and microbiological safety could be produced from mouldy sorghum. There is no literature available on value addition products to blackened sorghum except Shewale and Pandit (2009), also on value addition to sorghum through production of maltose syrup.

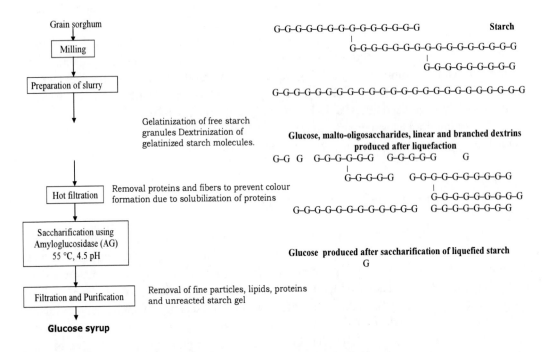

Figure 3. Process flow sheet for production of glucose syrup from sorghum.

Application of ultrasound is reported in wet milling process for isolation of starch from corn (Zhang et al., 2005), sorghum (Park et al., 2006) and rice (Wang and Wang 2004), and in dry corn milling ethanol production (Kinley et al., 2006; Khanal et al., 2007).

Normally isolated starch has been used to produce starch hydrolyzed products like maltodextrins, dextrose, glucose syrup, maltose and maltose syrup. Yields of starch isolation from sorghum were reported to be around 50–60% and from corn to be 60 - 70%. This means that the rest of the part (40 – 50% for sorghum and 30 – 40% for corn) gets wasted or does not fetch a respectable price. Whole grain flour can be used directly for liquefaction and saccharification rather than isolating starch. Such methodology of direct hydrolysis was first used by Kroyer in 1966 using corn grits for the production of glucose. Direct hydrolysis of flour of maize (Bos and Norr, 1974; Twisk et al., 1976), broken rice (Tegge and Ritcher, 1982) and sorghum (Tegge and Ritcher, 1982; Devarajan and Pandit, 1996; Aggarwal et al., 2003) was reported for the production of glucose. However there is no such literature available for production of maltose.

7.3.1. Value Addition to Sorghum through Production of Glucose Syrup

Recently, we, Shewale and Pandit (2009) worked on production of glucose from different qualities of sorghum viz. healthy, germinated and blackened. Production of glucose from sorghum flour consists of two reaction steps:

1. Liquefaction using bacterial α-amylase in which gelatinisation of free starch granules and dextrinization (depolymerisation) of gelatinized starch take place simultaneously. This produces mixture of malto-oligosaccharides, linear and branched dextrins.
2. Saccharification using amyloglucosidase (AG) in which AG cleaves first $\alpha(1 \to 4)$ linkage from non-reducing end glucose polymer and produces glucose.

Process of production of glucose syrup from sorghum flour with depiction of chemistry is given in Figure 3. Optimized reaction parameters for liquefaction and saccharification to produce glucose are given in Table 9.

Table 9. Optimized parameters for Liquefaction and Saccharification

Parameter	Liquefaction	Saccharification
Temperature (°C)	85	55
pH	6	4.5
Slurry concentration	30% w of sorghum flour / v of slurry	n.a.
BLA concentration	0.06% v/w of sorghum flour i.e. 0.086% v/w of sorghum starch	n.a.
AG concentration	n.a.	6 AGU/mL of liquefact i.e. 0.058 % v/ w of starch
$CaCl_2$ concentration	200 ppm	n.a.

Adapted from Shewale and Pandit, 2009.

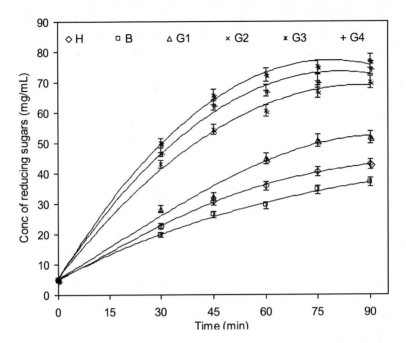

Figure 4. Effect of quality of sorghum on liquefaction. Reaction conditions: 30% w/v sorghum slurry, 85 °C, pH 6 and $CaCl_2$ concn = 200 ppm. Data points without shadow—blue color on starch–iodine reaction. Data points with shadow—disappearance of blue color (that is, dark red with tinge of violet) on starch–iodine reaction i.e. liquefaction complete. H—healthy sorghum; B—blackened sorghum; G1, G2, G3, G4—healthy sorghum germinated for 12 h, 24 h, 36 h, and 48 h, respectively, after steeping for 12 h in plain water. Adapted from Shewale and Pandit (2009).

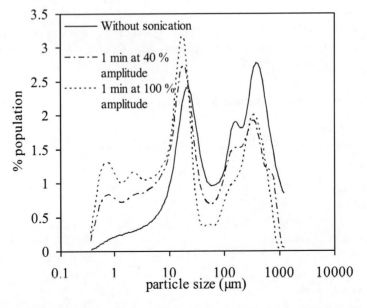

Figure 5. Effect of prior ultrasound treatment on the particle size distribution in 30% w/v sorghum flour slurry. Adapted from Shewale and Pandit (2009).

Shewale and Pandit (2009) has compared liquefaction of sorghum with different qualities viz. healthy, germinated and blackened sorghum. Shewale and Pandit reported that liquefaction progress of blackened sorghum is slightly slower than that of healthy sorghum (Figure 4) probably due to enzyme inhibition due to mycotoxins present in the blackened pericarp; however liquefaction of germinated sorghum is much faster as compared to counterparts of healthy sorghum (Figure 4) attributed to the loosening of the protein cage surrounding starch granules due to proteases produced (that is, protein matrix-degrading enzymes) during the process of germination.

Shewale and Pandit (2009) reported that integration of ultrasound pretreatment to sorghum slurry in the process prior to liquefaction increased the percentage of saccharification from 84 to 90 for healthy sorghum, from 85 to 90 for germinated sorghum and from 86 to 87 for blackened sorghum. They attributed the role of ultrasound treatment of the sorghum slurry prior to liquefaction in enhancing saccharification performance to disruption of the hydrophobic protein matrix surrounding the starch granules and the amylose–lipid complex due to physical effects of acoustic cavitation, like shock-wave propagation and microjet formation in the vicinity of a liquid–solid interface; followed by release of more starch granules (this can be observed with an increase in the peak area corresponding to the particle diameter of 10-20 μm (Figure 5); which is the diameter of sorghum starch granule according to Tester et al., 2004 a and b) and their availability for further action of α-amylase and amyloglucosidase. They also reported that if sorghum liquefact was directly taken for saccharification without the filtration step, percent values of saccharification were observed to be 87–89 and 93–95 without and with ultrasound treatment, respectively. As integration of short ultrasound treatment (about 1 min) in the production of glucose from dry milled sorghum increased glucose production and its possible subsequent use in the bioethanol production will result in an increase in the yield of ethanol, and hence improve the economic feasibility of the process of producing bioethanol from sorghum.

7.3.2. Value Addition to Sorghum through Production of Maltose Syrup

Production of maltose syrup from sorghum flour consists of two reaction steps:

1. Liquefaction using *B. licheniformis* α-amylase (BLA) in which gelatinisation of free starch granules and dextrinization (depolymerisation) of gelatinized starch take place simultaneously. This produces a mixture of malto-oligosaccharides, linear and branched dextrins.
2. Saccharification using barley β-amylase (BBA) with or without pullulanase (PL), in which BBA cleaves second α(1→4) linkage from non-reducing end of glucose polymer and produces maltose. Use of only BBA in saccharification will result into production of maltose and β limit dextrins, due to the inability of BBA to bypass α(1→6) linkages. If pullulanase is used along with BBA, pullulanase will cleave α(1→6) linkage and BBA can then attack the rest of the chain in the β limit dextrins. Hence saccharification using BBA and pullulanase will result into mostly maltose with small quantities of glucose and maltotriose.

Table 10. Optimized Parameters for Liquefaction and Saccharification

Parameter	Liquefaction	Saccharification
Temperature (°C)	85	50
pH	6	5.5
Slurry concentration	30% w of sorghum flour / v of slurry	n.a.
BLA concentration	0.06% v/w of sorghum flour i.e. 0.086% v/w of sorghum starch	n.a.
BBA concentration	n. a.	0.04 %v/ w of starch
PL concentration	n. a.	0.057 %v/w of starch
CaCl$_2$ concentration	200 ppm	n.a.

The process of the production of maltose syrup from sorghum flour, which is used in the present experimental work, is shown in Figure 6. Chemistry of the process has also been depicted from Figure 6.

Optimized conditions for liquefaction of sorghum flour and saccharification to maltose syrup are shown in Table 10.

According to our unpublished data, integration of ultrasound pretreatment to sorghum slurry in the process prior to liquefaction increased the percentage of saccharification to reducing sugars (maltose equiv) from 75 to 86 for healthy sorghum; from 79 to 86 for germinated sorghum and from 83 to 86 for blackened sorghum. Reasons behind increase in the percentage of saccharification due to ultrasound are the same as that discussed in earlier sections.

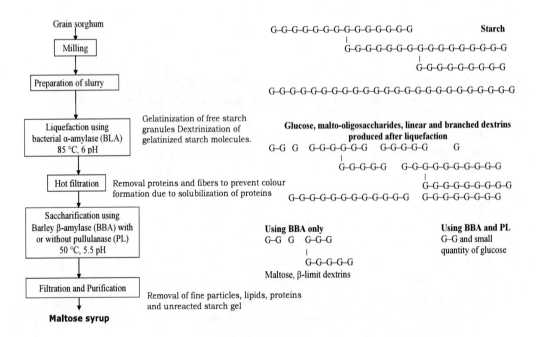

Figure 6. Process flow sheet for production of maltose syrup from sorghum.

7.3.3. Cost Economics of Value Addition to Sorghum

In the chemical project economics, cost of production plays a key role. It represents operating expenses, which are of a recurring nature. They have significant impact on the selling price and ultimately profitability. Operating expenses are incurred after the plant is commissioned and the production begins. There are mainly two parameters that are major and have direct impact on the cost of production viz. 1. Raw materials; 2. Utilities (Mahajani and Mokashi, 2005).

Cost for production of glucose from sorghum was determined by considering cost of raw materials and utilities only. Manpower costs are not considered. From Table 11, it can be seen that processing cost of production per kg of glucose produced from healthy, blackened sorghum is about 76% and 65% of the counterparts from isolated starch, respectively. This indicates that the production of glucose from sorghum is economically feasible.

Cost for production of maltose from sorghum was determined by considering cost of raw materials and utilities only. Manpower costs are not considered. From Table 12, it can be seen that cost of production per kg of maltose produced from healthy, blackened sorghum is about 80% and 66% of the counterparts from isolated starch, respectively. This indicates that the production of maltose from sorghum is economically feasible. Here it should be remembered that processing cost has been determined for production of dried maltose syrup. In order to obtain pure maltose, selective crystallization must be done and its cost has not been considered here.

Cost reported in Tables 11 and 12 are costs in India; but we believe that magnitude of difference between processing cost of production from isolated starch, healthy and blackened sorghum will remain the same.

7.3.4. Alternative Approaches for Value Addition to Sorghum

Possibility of value addition to healthy, blackened and germinated sorghum has been explored through the production of glucose and maltose syrup. However, there are other products also, that can be produced by using sorghum as a starting material. Flow sheet for the production of different products is provided in Figure 7. In fact industry can switch from one product to another depending upon the market needs.

The first step in the production of any product is liquefaction, i.e. simultaneous gelatinization of free starch granules and dextrinization of gelatinized starch using bacterial α-amylase. Liquefaction can be continued to achieve desired DE (15, 20 or 30). Then the sorghum liquefact needs to be filtered, purified and dried to get maltodextrins.

Glucose syrup (DE > 96) produced can further be processed using glucose isomerase to produce fructose syrup. Sorghum liquefact without any hot filtration can be saccharified using amyloglucosidase to glucose syrup with DE greater than 90. Then, this syrup can be fermented to produce ethanol. After completion of fermentation, ethanol is distilled out of this mixture. This product is termed as grain based alcohol or bioethanol. Remanent stillage can be dried to produce DDGS, which can be used as animal feed or can be anaerobically digested to produce bio-gas.

Table 11. Comparison of processing cost or cost production of glucose from sorghum

Parameter	Isolated starch	Healthy sorghum	Blackened sorghum	Germinated sorghum
Cost of Raw material	16.51	7.72	5	7.65
Cost of Utilities	4.58	6.69	6.69	6.69
Total cost	21.09	14.41	11.69	14.34
Glucose produced, kg	1.11	1	0.95	1
Processing cost, Rs/kg of glucose produced	19	14.41	12.3	14.34

Note: 100 % saccharification for pure starch (due to hydrolytic gain 1 kg starch produces 1.11 kg glucose);
90 % saccharification for health and germinated sorghum;
85 % saccharification for blackened sorghum.
Raw materials includes sorghum, water, enzymes. Utilities includes that required for liquefaction, saccharification, filtration and evaporation of water to produce 84 % syrup.

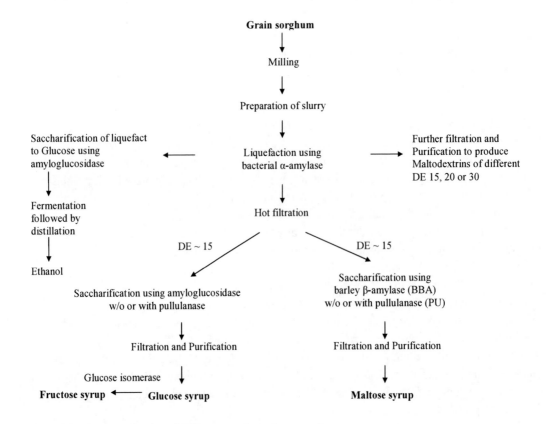

Figure 7. Production schemes of different products from sorghum.

Table 12. Comparison of processing cost or cost production of maltose syrup from sorghum

Parameter	Isolated starch	Healthy sorghum	Blackened sorghum	Germinated sorghum
Cost of Raw material	16.91	8.12	5.51	8.10
Cost of Utilities	4.38	6.41	6.41	6.41
Total cost	21.29	14.53	11.92	14.53
Maltose produced, kg	0.73	0.62	0.62	0.62
Reducing sugars produced, kg	1.05	0.89	0.89	0.89
Processing cost, Rs/kg of maltose produced	29.2	23.4	19.2	23.4
Rs/kg of reducing sugars produced	20.3	16.33	13.4	16.3

Note: 100 % saccharification for pure starch (due to hydrolytic gain 1 kg starch produces 1.05 kg maltose);
   85 % saccharification for healthy, blackened and germinated sorghum.
Reducing sugars (i.e. dry solids in maltose syrup) produced by using barley beta-amylase contains 70% maltose-assumption. Raw materials includes sorghum, water, enzymes Utilities includes that required for liquefaction, saccharification, filtration and evaporation of water to produce 84 % syrup.

8. Problem Areas and Factors Affecting them in Industrial Utilization of Sorghum

Sorghum has the distinct advantage (compared to other major cereals) of being drought resistant and many subsistence farmers in these regions cultivate sorghum as a staple food crop for consumption at home. Therefore sorghum acts as a principal source of energy, protein, vitamins and minerals for millions of the poorest people living in these regions. In this way, sorghum plays a crucial role in the world food economy as it contributes to rural household food security (Duodu et al., 2003). However, there are a few problem areas in the utilization of sorghum as food as well as industrial utilization and they are discussed here.

8.1. Gelatinization of Starch

Gelatinization of the starch is the most important initial step, due to which starch becomes more susceptible to enzyme action and completely digestible by starch hydrolyzing enzymes. Sorghum starch gelatinization temperature ranges were reported to be 67–73 °C and 71–81 °C for sorghums grown in southern Africa and in India, respectively (Taylor et al., 2006). Gelatinization temperature of sorghum starch is reported to be in the range of 75–80 °C (Palmer, 1992) and 60–80 °C (Wu et al., 2007).

Gelatinization temperature of sorghum starch is far higher than the range quoted for barley starch of 51–60 °C. This factor becomes important in the malting and brewing of sorghum. Due to this difference in the gelatinization temperature the simultaneous gelatinization and hydrolysis of starch that occurs when mashing barley malt, is problematic with sorghum malt. Sorghum grain or sorghum malt is first cooked to gelatinize the starch and then the starch is hydrolyzed using barley malt, commercial enzymes or a combination of the two.

Taylor et al. (2006) have reviewed the literature on starch gelatinization with the context of its use in malting and brewing. In a study of 30 sorghum varieties, Dufour et al. (1992) found a few with low gelatinisation temperatures, approaching that of barley. More recently, Beta et al. (2000a–c) found that Barnard Red, a traditional South African sorghum variety which was selected for its good malting and opaque beer characteristics, had a low onset starch gelatinization temperature of 59.4 °C and gave high paste viscosity, even though the starch had a normal amylose-amylopectin ratio. It is suggested that waxy sorghums gelatinize more rapidly, have a relatively weak endosperm protein matrix and are more susceptible to hydrolysis by amylases and proteases than normal endosperm sorghums and hence should be better for brewing (Del Pozo-Insfran et al., 2004). Figueroa et al. (1995) investigated mashing of 20 sorghum adjuncts of varying endosperm structure with barley malt. They found that the waxy and heterowaxy types gave much shorter conversion times (time to starch disappearance as indicated by iodine yellow colour) than normal types. They attributed this to the lower starch gelatinisation temperatures, 69.6 °C for waxy type, 71.1 °C for the heterowaxy type and 71.1–73.3 °C for the normal types. Interestingly, Ortega Villican and Serna-Saldivar (2004) found that when brewing with waxy sorghum adjunct, highest beer ethanol content and lowest residual sugar content were obtained if the adjunct was first heated at 80 °C, then pressure cooked.

Chandrashekar and Kirleis (1988) showed that the degree of starch gelatinization (using the β-amylase and pullulanase method) was lower in hard endosperm sorghum (with high kafirin protein content) than in soft endosperm types (low kafirin content). The addition of the reducing agent 2-mercaptoethanol during cooking markedly increased the degree of gelatinisation. However, increase in the degree of starch gelatinization was greater in harder high kafirin sorghum than that in the softer low kafirin sorghum. They concluded that the endosperm protein matrix which envelops the starch granules limits starch gelatinisation. They also reported that after treatment of pepsin with uncooked sorghum flour (soft sorghum), flour particles lose their structural integrity and only free starch granules with some adhering protein bodies remain indicating that the integrity of sorghum particles is maintained by protein. Their SEM work showed that flour particles from the hard grains were most often covered with a cell wall and when exposed the starch granules seemed to be surrounded by numerous protein bodies. In contrast, in the soft grains, cell walls appeared sloughed off the particle and far fewer protein bodies surrounding the starch granule. Starch granules, protein bodies and cell wall appear to be linked together by strands of protein. This gets supported by the fact that organized structure in both hard and soft grains was lost due to treatment with pepsin. They suggested that the linking proteins that hold the particle together are strands of glutelin. Thus, the protein matrix in the hard grain contains both protein bodies and matrix strands, whereas soft grains contains a large amount of strand protein.

Table 13. Effect of the waxy gene of kafir and pronase treatment on *in vitro* starch hydrolysis (mg glucose/g starch) using amyloglucosidase. Lichtenwalner et al. (1978)

Genotype	Amylose content %	Ground grain	Pronase pretreated ground grain	Purified starch
Normal (WxWxWx)	24	434	540	550
Heterowaxy (WxWxwx)	23.1	46	565	598
Heterowaxy (Wxwxwx)	17.3	545	703	745
Waxy (wxwxwx)	1	741	966	1035

8.2. Protein Digestibility

Sorghum plays a crucial role in the world food economy as it contributes to rural household food security (ICRISAT/ FAO). A nutritional constraint to the use of sorghum as food is the poor digestibility of sorghum proteins on cooking. Digestibility may be used as an indicator of protein availability. It is essentially a measure of the susceptibility of a protein to proteolysis. A protein with high digestibility is potentially of better nutritional value than one of low digestibility because it would provide more amino acids for absorption on proteolysis. *In vivo* studies using pepsin and *in vitro* studies show that the proteins of wet cooked sorghum are significantly less digestible than the proteins of other similarly cooked cereals like wheat and maize (Duodu et al., 2003). In an excellent review on factors affecting sorghum protein digestibility, Duodu et al. (2003) divided these factors into two broad categories:

Exogenous factors: These refer to factors that arise out of the interaction of sorghum proteins with non-protein components like polyphenols, non-starch polysaccharides, starch, phylates and lipids.

Endogenous factors: These refer to factors that arise out of changes within the sorghum proteins themselves and do not involve interaction of the proteins with non-protein components.

Duodu et al. (2003) have discussed all these factors in detail in the review. Since protein digestibility is not a topic of interest in the present work, this will not be discussed in detail here.

8.3. Starch Digestibility

Starch digestibility is an important parameter from the context of feeding value of sorghum as well as production of ethanol from sorghum. Sorghum flour with high starch digestibility will have good food value as well as will be a good candidate for production of ethanol.

Lichtenwalner et al. (1978) reported that amylose content decreased and *in vitro* starch digestibility using amyloglucosidase increased with incremental increases of the waxy gene in sorghum. Due to pronase treatment of the sorghum flour, starch digestibility increases in all four types of sorghum and slightly lesser than that of isolated sorghum starch. The pronase treatment significantly increased the rate of starch hydrolysis because it hydrolyzed the protein matrix (which surrounds starch granules) and increased the surface area of the starch in contact with amyloglucosidase. Their data is shown in Table 13.

Rooney and Pflugfelder (1986) have reviewed factors affecting starch digestibility with special emphasis on sorghum and corn. The digestibility of starch is affected by the composition and physical form of the starch, protein-starch interactions, the cellular integrity of the starch-containing units, antinutritional factors and the physical form of the feed or food material. Starch exists inside the endosperm of cereals enmeshed in a protein matrix, which is particularly strong in sorghum (Figure 8). Starch digestibility is affected by the plant species, the extent of starch-protein interaction, physical form of the granule, inhibitors such as tannins, and the type of starch. Among the cereals, sorghum generally has the lowest raw starch digestibility due to restrictions in accessibility to starch caused by endosperm proteins.

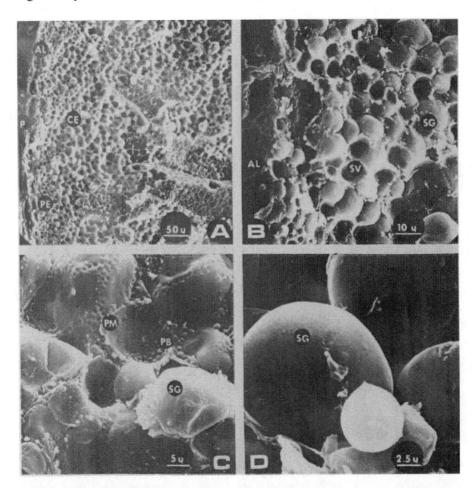

Figure 8. SEM images of intermediate texture sorghum kernels. A) Endosperm cross section (P = pericarp, AL = aleurone cell layer, PE = peripheral endosperm, CE = corneous endosperm; approx 200X). B) Corneous endosperm area (SV = starch void, SG = starch granule; approx 1,000X). C) Protein and starch of corneous endosperm (PM = protein matrix, PB = protein bodies, SG = starch granule; approx 2,000X). D) Starch of floury endosperm (approx 4,000X). (Source: Rooney and Pflugfelder, 1986)

High-amylose corn (amylomaize) has poor digestibility in both raw and cooked forms, while waxy cereal starches are among the most digestible of all starches. Digestibility of a starch is generally inversely proportional to amylose content i.e. directly proportional to waxyness of the sorghum. Rooney and Pflugfelder, 1986 have also explained briefly the reasons behind lower digestibility of sorghum as compared to corn.

Zhang and Hamaker (1998) found that digestibility (using porcine pancreatic α-amylase) of cooked isolated sorghum starches was markedly higher than starch from cooked sorghum flours. They observed that pepsin pretreatment before cooking increased the starch digestibility of sorghum flour by 7–14%, but no significant increase in starch digestibility was seen when pepsin treatment was performed after cooking. These authors also reported that after cooking with a reducing agent, 100 mM sodium metabisulfite, starch digestibility of sorghum flours increased significantly. Elkhalifa et al. (1999) also observed that *in vitro* starch digestibility (IVSD) of the treated gruel initially increased in the presence of cysteine, sodium metabisulphite or ascorbic acid; however, at high levels of cysteine or sodium metabisulphite the IVSD was low. Ezeogu et al. (2005) reported that starch digestibility (using pancreatic porcine α-amylase) was significantly higher in floury sorghum endosperm than vitreous endosperm and cooking with reducing agent, 2-mercaptoethanol, increased starch digestibility in sorghum, and more with vitreous endosperm flours.

The fact that reducing agents improved sorghum starch digestibility suggests that disulphide bond cross-linking within the kafirin-containing endosperm protein matrix is responsible for the reduced gelatinisation in sorghum. This is the same mechanism that has been implicated in the reduced protein digestibility of cooked sorghum (Duodu et al., 2003). This interpretation is supported by the work of Ezeogu et al. (2005) who found evidence of disulphide bond cross-linked prolamin proteins in high proportion and extensive polymerization through disulphide bonding of prolamins on cooking of sorghum through SDS-PAGE, with the formation of high molecular weight polymers (M. Wt > 100k). Also formation of web-like or sheet-like protein structures due to a disulphide-mediated protein polymerization process during mashing or heat moisture treatment is reported by Hamaker and Bugusu (2003) and Wu et al. (2007). Ezeogu et al. (2005) also observed that pressure cooking the flours improved starch digestibility of vitreous (hard) and floury (soft) endosperm maize and sorghum flours and markedly so for sorghum vitreous endosperm flour. They suggested that pressure cooking could have physically disrupted the protein matrix.

8.4. Tannin Content in Sorghum

Tannin is located in the testa portion of sorghum grain. Tannins confer valuable agronomic properties on sorghum, including protection against insects, birds and weather damage. However, tannins inactivate extracted malt amylases (Beta et al., 2000a–c; Daiber, 1975), significantly reducing starch breakdown and sugar production during brewing (Daiber, 1975). Tannins are well known for their adverse effect on starch digestibility because of their ability to interact with proteins (including hydrolytic enzymes), metal ions, and polysaccharides (Wu et al., 2007). Wu et al. (2007) found that the liquefaction of starch in tannin sorghums was more difficult and slower than in normal and waxy sorghums. Wu et al. (2007) also confirmed that tannin contents had a strong adverse effect on conversion efficiency of sorghum to ethanol. Taylor et al. (2006) have reviewed the use of tannin sorghum in malting and brewing.

REFERENCES

Aggarwal N. K.; Nigam P.; Singh D.; Yadav B. S., Process optimization for the production of sugar for the bioethanol industry from sorghum, a non-conventional source of starch. *World Journal of Microbiology and Biotechnology,* 2001, 17, 411-415.

Agu, R.C., Palmer, G.H. A reassessment of sorghum for lager-beer brewing. *Bioresource Technology,* 1998, 66, 253–261.

Ahmed KM and Ravinder Reddy Ch. 1993. A pictorial guide to the identification of seedborne fungi of sorghum, pearl millet, chickpea, pigeonpea and groundnut. Information Bulletin No. 34. Patancheru, Andhra Pradesh 502 324, India: ICRISAT. 200 pp.

Aisen, A. O.; Palmer, G. H.; Stark, J. R.; The Development of Enzymes During Germination and Seedling Growth in Nigerian Sorghum. *Starch,* 1983, 35, 316-320.

Anglani, C. Sorghum for human food: a review. *Plant Foods Hum. Nutr.*, 1998, 52, 85-89.

Awika, J.M., McDonough, C.M., Rooney, L.W. Decorticating sorghum to concentrate healthy phytochemicals. J. of Agricultural and Food Chemistry, 2005, 53, 6230–6234.

Bandyopadhyay, R.; Butler, D. R.; Chandrashekar. A.; Reddy, R. K. and Navi, S. S. Technical and institutional options for sorghum grain mold management: proceedings of an international consultation, 18-19 May 2000, ICRISAT, Patancheru, India. (Chandrashekar, A., Bandyopadhyay, R., and Hall, A.J., eds.).

Belton, P.S.; Delgadillo, I.; Halford, N.G.; Shewry, P.R. Kafirin structure and functionality. *Journal of Cereal Science,* 2006, 44, 272–286.

Beta, T., Corke, H., Rooney, L.W., Taylor, J.R.N. Starch properties as affected by sorghum grain chemistry. *Journal of the Science of Food and Agriculture,* 2000b, 81, 245–251.

Beta, T., Corke, H., Taylor, J.R.N. Starch properties of Barnard Red, a South African red sorghum of significance in traditional African Brewing. *Starch,* 2000a, 52, 467–470.

Beta, T., Rooney, L.W., Marovatsanga, L.T., Taylor, J.R.N. Effect of chemical treatments on polyphenols and malt quality in sorghum. J. of Cereal Science, 2000c, 31, 295–302.

Bos, C. and Norr, N. J. Experiences with the DDS-Kroyer direct hydrolysis process. *Starch,* 1974, 26, 181-185.

Buffo, R. A., Weller, C.L.,Gennadios, A. Films from Laboratory-Extracted Sorghum Kafirin. Cereal chemistry, 1997, 74, 473–475.

Buffo, R.A., Weller, C.L., Parkhurst, A.M. Optimization of sulfur dioxide and lactic acid steeping concentrations for wet-milling grain sorghum. *Transactions of the American Society of Agricultural Engineers,* 1997, 40, 1643–1648.

Buffo, R.A., Weller, C.L., Parkhurst, A.M. Wet-milling factors of sorghum and relationship to grain quality. *Journal of Cereal Science,* 1998, 27, 327–334.

Carson, L., Setser, C., Sun, X. S. Sensory characteristics of sorghum composite bread. Int. J. Food Sci. Technol., 2000, 35, 465-471.

Chandrashekar, A. and Kirleis, A. W. Influence of protein on starch gelatinization in sorghum. *Cereal Chem.*, 1988, 65, 457-462.

Chandrashekar, A.; Satyanarayana, K.V. Disease and pest resistance in grains of sorghum and millets. *Journal of Cereal Science*, 2006, 44, 287–304.

Chandrashekhar, A.; Mazhar, H. The Biochemical Basis and Implications of Grain Strength in Sorghum and Maize. *Journal of Cereal Science*, 1999, 30, 193-207.

Corredor, D.Y.; Bean, S.R.; Schober, T.; Wang, D. Effect of decorticating sorghum on ethanol production and composition of DDGS. *Cereal Chemistry*, 2006, 83, 17–21.

Da Silva, L. S, Taylor, J. R. N. Physical, mechanical, and barrier properties of kafirin films from red and white sorghum milling fractions. Cereal Chemistry, 2005, 82, 9-14.

Dahlberg, J. A. Classification and characterization of sorghum in *Sorghum: Origin, History, Technology, and Production.* Smith, C, W.; Frederiksen, R. A. Eds. Wiley, New york, 2000.

Daiber, K.H. Enzyme inhibition by polyphenols of sorghum grain and malt. *Journal of the Science of Food and Agriculture*, 1975, 26, 1399–1411.

Deb, U. K.; Bantilan, M. C. S.; Ro, A. D.; Rao P. R. Global Sorghum Production Scenario in *Sorghum genetic enhancement: research process, dissemination and impacts.* Bantilan, M. C. S.; Deb, U. K.; Gowda, C. L. L.; Reddy, B. V. S.; Obilana, A. B. and Evenson, R. E. Eds. 2004. International Crops Research Institute for the Semi-Arid Tropics (ICRSAT).

Del Pozo-Insfran, D., Urias-Lugo, D., Hernandez-Brenes, C., Serna Saldivar, S.O. Effect of amyloglucosidase on wort composition and fermentable carbohydrate depletion in lager beers. *Journal of the Institute of Brewing*, 2004, 110, 124–132.

Dendy, D. A. V. Sorghum and Millets Chemistry and Technology. U.S.A., St. Paul, Minnesota, American Association of Cereal Chemists, Inc., 1995, pp. 406.

Devarajan B.; Pandit A. B., Sorghum flour as Raw Material for Glucose Production. *J. Maharashtra Agric. Univ.*, 1996, 21 (1), pg. 86-90.

Diao, Y., Walawender, W., Fan, L. Activated carbons prepared from phosphoric acid activation of grain sorghum. *Biosource Technology*, 2002, 81, 45–52.

Dicko, M. H., Gruppen H., Traoré A. S., Voragen, A. G. J. and van Berkel, W. J. H. Sorghum grain as human food in Africa: relevance of content of starch and amylase activities. *African Journal of Biotechnology*, 2006, 5, 384-395.

Dufour, J.P.; Melotte; L., Srebrnik, S. Sorghum malts for the production of a lager beer. *Journal of the American Society of Brewing Chemists*, 1992, 111, 110–119.

Duodu, K.G.; Taylor, J.R.N.; Belton, P.S.; Hamaker, B.R. Factors affecting sorghum protein digestibility. *Journal of Cereal Science*, 2003, 38, 117–131.

Dykes, L. and Rooney, L. W. Sorghum and millet phenols and antioxidants. *Journal of Cereal Science*, 2006, 44, 236–251.

Elkhalifa, A. O.; Chandrashekar, A.; Mohamedc, B.E.; Tinay, A.H. Effect of reducing agents on the in vitro protein and starch digestibilities of cooked sorghum. *Food Chemistry*, 1999, 66, 323-326.

Emmambux, M. N., Stading, S., Taylor, J. R. N. Tannic acid modifies material properties of sorghum kafirin film. Annual transactions of the Nordic rheology society, 2004, vol 12., 251-254.

Ezeogu, L.I.; Duodua, K.G.; Taylora, J.R.N. Effects of endosperm texture and cooking conditions on the in vitro starch digestibility of sorghum and maize flours. *Journal of Cereal Science*, 2005, 42, 33–44.

Figueroa, J.D.C.; Martinez, B.F.; Rios, E. Effect of sorghum endosperm type on the quality of adjuncts for the brewing industry. *Journal of the American Society of Brewing Chemists*, 1995, 53, 5–9.

Frederiksen, R. A. Diseases and Disease management in sorghum in *Sorghum: Origin, History, Technology, and Production.* Smith, C, W.; Frederiksen, R. A. Eds. Wiley, New york, 2000.

Fredrick, E. J. Effect of sorghum flour composition and particle size on quality of gluten-free bread. Thesis for M. S., Kansas state university, 2009.

Hallgren, L. Lager beers from sorghum. in *Sorghum and Millets: Chemistry and Technology*. Dendy, D.A.V. Ed. American Association of Cereal Chemists, St. Paul, MN, U.S.A., 1995, pp. 283–297.

Hamaker, B. R., and Bugusu, B. A. Overview: Sorghum proteins and food quality. In: Proc. AFRIPRO Workshop on the Proteins of Sorghum and Millets: Enhancing Nutritional and Functional Properties for Africa. 2003, P. S. Belton and J. R. N. Taylor, eds. Available at http://www.afripro.org.uk/papers/Paper08Hamaker.pdf. Pretoria, South Africa.

Harlan, J. R. and de Wet, J.M.J. A simplified classification of cultivated sorghum. *Crop Science,* 1972, 12, 172-176.

Higiro, J., Flores, R.A., Seib, P.A. Starch production from sorghum grits. *Journal of Cereal Science*, 2003, 37, 101–109.

Hilhorst, R. Bierbereiding in Burkina Faso. PtIProcestechniek, 1986, 41, 93-95.

Hizukuri, S. Polymodal distribution of the chain lengths of amylopectin and its significance. *Carbohydrate Research*, 1986, 147, 342–347.

Hugo, L. F., Rooney, L. W., Taylor, J. R. N. Fermented sorghum as a functional ingredient in composite breads. Cereal Chemistry, 2003, 80: 495-499.

Hugo, L. F., Rooney, L. W., Taylor, J. R. N. Malted sorghum as a functional ingredient in composite bread. Cereal Chemistry, 2000, 77, 428-432.

International Crops Research Institute for the Semi-Arid Tropics (ICRISAT)/Food and Agriculture Organization (FAO), 1996. The World Sorghum and Millet Economies. Facts, Trends and Outlook. ICRISAT, Patancheru, India/ FAO, Rome.

Jeyakumar S. Sorghum demand on the rise, Facts for you, 2010 (September), 14-16.

Kearsley, M. W. and Dziedzic, S. Z. `Handbook of Starch Hydrolysis Products and Their Derivatives' Blackie Academic and Professional, London, 1995.

Khanal, S.; Montalbo, M.; Leeuwen, J.; Srinivasan, G.; Grewell, D. Ultrasound Enhanced Glucose Release From Corn in Ethanol Plants. *Biotechnology and Bioengineering*, 2007, 98, 978-985.

Kimber, C. T. Origins of domesticated sorghum and its early diffusion to India and China in *Sorghum: Origin, History, Technology, and Production.* Smith, C, W.; Frederiksen, R. A. Eds. Wiley, New york, 2000.

Kinley, M. T.; Snodgrass, J. D. and Krohn, B. Alcohol production using sorghum. US patent 7101691, 2006.

Kleih, U.; Ravi, S. B.; Rao, B. D. and Yoganand, B. Industrial Utilization of Sorghum in India. Ejournal.icrisat.org, 3, 2007. This study is based upon fieldwork undertaken in mid-1998 in the context of the project 'Sorghum in India: Technical, policy, economic, and social factors affecting improved utilization', which was funded by the Department for International Development (DFID) and jointly undertaken by ICRISAT, NRCS, and NRI.

Kroyer, K. *Staerke*, 1966, 10, 312.

Léder, I. SORGHUM AND MILLETS, in *Cultivated Plants, Primarily as Food Sources*, Füleky, G., in *Encyclopedia of Life Support Systems (EOLSS)*, Developed under the Auspices of the UNESCO, Eolss Publishers, Oxford ,UK.

Lichtenwalner, R. E.; Ellis, E. B. and Rooney L. W. Effect of Incremental Dosages of the Waxy Gene of Sorghum on Digestibility. *J Anim Sci.*, 1978, 46, 1113-1119.

Mahajani, V. V.; Mokashi, S. M. Chemical Projects Economics. Macmillan India Ltd. 2005. pp. 150-160.

Mezo-Villanueva, M. and Serna-Saldivar, S. O. Effect of protease addition on starch. recovery from steeped sorghum and maize. *Starch*, 2004, 56, 371-378.

Mishra, S.K. Replacement of yellow maize with tannin-free sorghum in white leghorn layer diet Indian Journal of Poultry Science, 1995, 30, 76–78.

Mojović, L, Pejin, D., Grujić, O., Markov, S., Pejin, J., Rakin, M., Vukašinović, M., Nikolić, S., Savic, D. Progress in the production of bioethanol on starch based feedstocks. Chemical Industry and Chemical Engineering Quarterly, 2009, 15, 211−226.

Munck, L. New milling technologies and products: Whole plant utilization by milling and separation of the botanical and chemical components in *Sorghum and Millets: Chemistry and Technology*, Dendy, D.A.V. (Ed.). American Association of Cereal Chemists, St. Paul, MN, U.S.A., 1995, pp. 223–281.

Murty, D.S., Kumar, K.A. Traditional uses of sorghum and millets in *Sorghum and Millets: Chemistry and Technology*, Dendy, D.A.V. (Ed.). American Association of Cereal Chemists, St. Paul, MN, U.S.A., 1995, pp. 185–221.

Musigakun, P. and Thongngam, M., Characteristics and functional properties of sorghum protein. *Kasetsart J. (Nat. Sci.)*, 2007, 41, 313 – 318.

Ortega Villicana, M.T.; Serna-Saldivar, S.O. Production of lager from sorghum malt and waxy grits. *Journal of the American Society of Brewing Chemists*, 2004, 62, 131–139.

Owuama, C.I. Brewing beer with sorghum. *J. of Institute of Brewing*, 1999, 105, 23–34.

Owuama, C.I. Sorghum: a cereal with lager beer brewing potential. World Journal of *Microbiology and Biotechnology*, 1997, 13, 253–260.

Palmer, G. H. Review: Sorghum – Food, Beverage and Brewing potential. *Process Biochemistry*, 1992, 27, 145-153.

Park, S. H.; Bean, S. R.; Wilson, J. D. and Schober, T. J. Rapid Isolation of Sorghum and Other Cereal Starches Using Sonication. *Cereal Chemistry*, 2006, 83, 611-616.

Perez Siraa, E. E.; Amaiz, M. L. A laboratory scale method for isolation of starch from pigmented sorghum. *Journal of Food Engineering*, 2004, 64, 515–519

Perten H. Practical experience in processing and use of millet and sorghum in Senegal and Sudan, *Cereal Foods World*, 1983, 28, 680-683.

Rama Rao, S.V., Praharaj, N.K., Raju, M.V.L.N., Mohapatra, S.C., Chawa, M.M., and Rooney, L. W. and Pflugfelder, R. L. Factors affecting starch digestibility with special emphasis on sorghum and corn. *J. Anim. Sci.*, 1986, 63, 1607-1623.

Rooney, L.W., Serna-Saldivar, S.O. Sorghum in Handbook of Cereal Science and Technology, Kulp, K.; Ponte, Jr., J.G. (Eds.), second ed. Marcel Dekker, New York, 2000, pp. 149–175.

Rooney, L.W., Waniska, R.D. Sorghum food and industrial utilization. in *Sorghum: Origin, History, Technology, and Production*, Smith, C.W., Frederiksen, R.A. (Eds.) Wiley, New York, 2000, pp. 689–729.

Schenck, F. W. and Hebeda, R. E. *Starch Hydrolysis Products: Worldwide Technology, Production, and Applications*, VCH Publishers 1992, New York.

Schober, T. J., Messerschmidt, M., Bean, S. R., Park, S. H., and Arendt, E. K. Gluten-free bread from sorghum: quality differences among hybrids. Cereal Chemistry, 2005, 82, 394-404.

Schober TJ, Bean SR, Arendt EK, Fenster C. Use of sorghum flour in bakery products. AIB International Technical Bulletin, 2006, 28.

Schober, T. J., Bean, S. R., Boyle, D. L. Gluten-free sorghum bread improved by sourdough fermentation: Biochemical, rheological, and microstructural background. *J Agric Food Chem.*, 2007, 55, 5137-5146.

Sheorain V, Banka R, and Chavan M., Ethanol production from sorghum. Technical and institutional options for sorghum grain mold management: proceedings of an international consultation, 18-19 May 2000, ICRISAT, Patancheru, India (Chandrashekar, A., Bandyopadhyay, R., and Hall, A.J., eds.). Patancheru 502 324, Andhra Pradesh, India: International Crops Research Institute for the Semi-Arid Tropics, Pages 228-239.

Shewale, S. D. and Pandit, A. B. Enzymatic production of glucose from different qualities of grain sorghum and application of ultrasound to enhance the yield. *Carbohydrate Research*, 2009, 344, 52–60.

Singh, V., Moreau, R.A., Hicks, K.B. Yield and phytosterol composition of oil extracted from grain sorghum and its wet-milled fractions. *Cereal Chemistry*, 2003, 80, 126–129.

Siruguri, V., Ganguly, C., Bhat, R. V. Utilization of mouldy sorghum and Cassia tora through fermentation for feed purposes. *African Journal of Biotechnology*, 2009, 8, 6349-6354.

Stahlman, P. W.; Wicks, G. A. Weeds and their control in grain sorghum in *Sorghum: Origin, History, Technology, and Production*. Smith, C, W.; Frederiksen, R. A. Eds. Wiley, New york, 2000.

Subramanium, V., and Metta, V. C. Sorghum grain for poultry feed. Technical and institutional options for sorghum grain mold management: proceedings of an international consultation, 18-19 May 2000, ICRISAT, Patancheru, India (Chandrashekar, A., Bandyopadhyay, R., and Hall, A.J., eds.). Patancheru 502 324, Andhra Pradesh, India: International Crops Research Institute for the Semi-Arid Tropics. Pages 242 - 247.

Suresh, K.; Kiran sree, N. and Venkateswer Rao, L. Production of ethanol b raw starch hydrolysis and fermentation of damaged grains of wheat and sorghum. *Bioprocess engineering*, 1999b, 21, 165-168.

Suresh, K.; Kiran sree, N. and Venkateswer Rao, L. Utilization of damaged sorghum and rice grains for ethanol production by simultaneous saccharification and fermentation. *Bioresource Technology*, 1999a, 68, 301-304.

Taylor, J. and Taylor, J. R. N. Some potential applications for brewers spent grain the protein rich co-product from sorghum lager beer brewing. The institute of brewing and distilling Africa sec. 12th scientific and technical convention (2009).

Taylor, J. R. N.; Schober, T. J.; Bean, S. R. Novel food and non-food uses for sorghum and millets. *Journal of Cereal Science*, 2006, 44, 252–271.

Taylor, J., Taylor, J. R. N., Dutton, M. F., De Kock, S., Identification of kafirrin film casting solvents. *Food chemistry*, 2005, 32, 149-154.

Taylor, J.R.N., Dewar, J. Fermented products: Beverages and porridges in *Sorghum: Origin, History, Technology, and Production*, Smith, C.W., Frederiksen, R.A. (Eds.). Wiley, New York, 2000, pp. 751–795.

Taylor, J.R.N.; Dewar, J. Developments in sorghum food technologies in *Advances in Food and Nutrition Research*, vol. 43, Taylor, S. (Ed.). Academic Press, San Diego, CA, U.S.A., 2001, pp. 217–264.

Teetes, G. L.; Pendleton, B. B. Insect pests of sorghum in *Sorghum: Origin, History, Technology, and Production.* Smith, C, W.; Frederiksen, R. A. Eds. Wiley, New york, 2000.

Tegge, V. G. and Richter G. Sorghum and broken rice as basic materials for glucose production. *Starch*, 1982, 34, 386-390.

Tester, R. F.; Karkalas, J.; Qi, X. Starch structure and digestibility Enzme-Substrate relationship. *World's Poultry Science Journal*, 2004b, 60, 186-195.

Tester, R. F.; Karkalas, J.; Qi, X. Starch—composition, fine structure and architecture. A Review *Journal of Cereal Science*, 2004a, 39, 151–165.

Thakur, R. P., Reddy, B. V. S., Indira, S., Rao, V. P., Navi, S. S., Yang, X. B. and Ramesh, S. 2006. Sorghum Grain Mold. Information Bulletin No. 72. Patancheru 502 324, Andhra Pradesh, India: International Crops Research Institute for the Semi-Arid Tropics. 32pp.

Thakur, R.S., Gupta, P.C., and Lodhi, G.P. Feeding value of different varieties of sorghum in broiler ration. Indian Journal of Poultry Science, 1984, 19, 103–107.

Trezza, T. A. and Vergano, P. J. Grease resistance of corn zein coated paper. *J. Food Sci.*, 1994, 59, 912-915.

Twisk, P. V.; Meltze, B. W. And Cormack R. H. Production of glucose from maize grits on commercial scale. *Starch*, 1976, 28, 23-25.

Wang, D., Bean, S., McLaren, J., Seib, P., Madl, R., Tuinstra, M., Shi, Y., Lenz, M. Wu, X., Zhao, R. Grain sorghum is a viable feedstock for ethanol production. *J Ind Microbiol Biotechnol*, 2008, 35, 313–320.

Wang, F.C., Chung, D.S., Seib, P.A., Kim, Y.S. Optimum steeping process for wet milling of sorghum. *Cereal Chemistry*, 2000, 77, 478–483.

Wang, L.; Wang, Y. Application of High-Intensity Ultrasound and Surfactants in Rice Starch Isolation. *Cereal Chemistry*, 2004, 81, 140-144.

Waniska, R. D. and Rooney, L. W. Structure and Chemistry of sorghum Caryopsis in *Sorghum: Origin, History, Technology, and Production.* Smith, C, W.; Frederiksen, R. A. Eds. Wiley, New york, 2000.

Wu, X.; Zhao, R.; Bean, S. R.; Seib, P. A.; Mclaren, J. S.; Madl, R. L.; Tuinstra, M.; Lenz, M. C. and Wang, D. Factors Impacting Ethanol Production from Grain Sorghum in the Dry-Grind Process. *Cereal Chemistry*, 2007, 84, 130-136.

Xie, X.J. and Seib, P.A. Laboratory wet-milling of grain sorghum with abbreviated steeping to give two products. *Starch/Staerke*, 2002, 54, 169–178.

Xie, X.J., Liang, Y.T.S., Seib, P.A., Tuinstra, M.R. Wet-milling of grain sorghum of varying seed size without steeping. *Starch/Staerke*, 2006, 58, 353–359.

Yang, P. and Sieb, P. Low-input wet-milling of grain sorghum for readily accessible starch and animal feed. *Cereal Chem.*, 1996, 73, 751–755.

Yang, R., Seib, P.A. Low-input wet-milling of grain sorghum for readily accessible starch and animal feed. *Cereal Chemistry*, 1995, 72, 498–503.

Zhan, X.; Wang, D.; Tuinstra, M.R.; Bean, S.; Seib, P.A. and Sun, X.S. Ethanol and lactic acid production as affected by sorghum genotype and location. *Industrial Crops and Products*, 2003, 18, 245-255.

Zhan, X.; Wanga, D.; Beanb, S.R.; Moc, X.,; Sunc, X.S. and Boyled, D. Ethanol production from supercritical-fluid-extrusion cooked sorghum. *Industrial Crops and Products*, 2006, 23, 304–310.

Zhang, G. and Hamaker, B. R. Low a-Amylase Starch Digestibility of Cooked Sorghum Flours and the Effect of Protein. *Cereal Chem.*, 1998, 75, 710-713.

Zhang, Z.; Niu, Y.; Exkhoff, S. R. and Feng, H. Sonication enhanced corn starch separation. *Starch*, 2005, 57, 240-245.

Zhao, R.; Bean, S.R.; Ioerger, B. P.; Wang, D.; Boyle, D. L. Impact of Mashing on Sorghum Proteins and Its Relationship to Ethanol Fermentation. *J. Agric. Food Chem.*, *2008*, 56, 946–953.

In: Sorghum: Cultivation, Varieties and Uses
Editor: Tomás D. Pereira

ISBN 978-1-61209-688-9
© 2011 Nova Science Publishers, Inc.

Chapter 2

BIO-FUEL AGRO-INDUSTRIAL PRODUCTION SYSTEM IN SWEET SORGHUM

Ramesh C. Ray[1,] and Shuvashish Behera[2]*

[1]Central Tuber Crops Research Institute, Bhubaneswar 751 019, Orissa, India
[2]Department of Botany, Utkal University, Vanivihar,
Bhubaneswar 751 004, Orissa, India

ABSTRACT

The consumption of bio-ethanol as bio-fuel may reduce greenhouse gases and gasoline imports. Also it can be replaced with lead or MTBE (Methyl tert-butyl ether) that are air or underground water pollutants, respectively. Plants are the best choice for meeting the projected bio-fuel demands. Recent studies have revealed that sweet sorghum can be used as a feedstock for bio-ethanol and bio-diesel production under hot and dry climatic conditions because it has a higher tolerance to salt and drought compared to sugarcane, sugar beet, maize and cassava that are currently used for bio-fuel production in the world. In addition, the high carbohydrate contents of sweet sorghum stalk are similar to sugarcane but its water and fertilizer requirements are much lower than sugarcane. The recent developments such as solid state fermentation, simultaneous saccharification and fermentation technology, use of mixed culture, recombinant yeast, and immobilized techniques, and the factors such as pH, agitation, particle stuffing rate, etc. in bio-fuel production from sweet sorghum are been discussed in this chapter.

INTRODUCTION

The depletion of fossil fuel reserves, the unstable panorama of the petroleum product prices and more recently, increasing environmental and political pressures have increased industrial focus toward alternative fuel resources and encouraged the search of products

---

[*] Corresponding author: Tel/Fax: 91-674-2470528: E-mail: rc_rayctcri@rediffmail.com

originating from plant biomass, as renewable sources of energy (Behera *et al.*, 2010a, b). Plants such as sugarcane (*Saccharum officinarum* L.), sugar beet (*Beta vulgaris* L.), maize (*Zea mays* L.), cassava (*Manihot esculenta* Crantz) and sweet sorghum (*Sorghum bicolor* L.), are the best choice for meeting the projected bio-ethanol demand. Sweet sorghum has an advantage over other bio-ethanol crops because of its adaptability over a wide range of climatic conditions, is least affected by climatic changes and has a higher benefit: cost ratio of cultivation and comparable ethanol yield per hectare of cultivation (Nadir *et al.*, 2009). Sweet sorghum is similar to grain sorghum but with sugar-rich stalks. It is a multi-purpose crop which can be cultivated for simultaneous production of grains from its ear-head as food and feed ingredients, sugary juice from its stalks for making syrup, bio-ethanol and bio-diesel, and bagasse and green foliage as an excellent fodder for animals, as organic fertilizer, or for paper manufacturing (Reddy *et al.*, 2007). Sweet sorghum produces grain 3-7 tons/ha and stalk 54-69 tons/ha (Almodares and Hadi, 2009). The sugar content in the juice of sweet sorghum varies in different varieties and the Brix ranges generally 16-22% (Almodares and Sepahi, 1996; Reddy *et al.*, 2007). Besides having rapid growth, high sugar accumulation and biomass production potential, sweet sorghum has wider adaptability in regard to soil types, agro-climatic conditions and rainfall patterns (Reddy *et al.*, 2005). Further, this crop has several other advantages such as drought resistance (Tesso *et al.*, 2005), water logging resistance (Reddy *et al.*, 2007), salinity resistance (Almodares *et al.*, 2007, 2008a) and is less affected by pests and diseases. Above all, sweet sorghum is a C4 crop with high photosynthetic efficiency (Reddy *et al.*, 2007).

2. SWEET SORGHUM AGRONOMY

Sweet sorghum cultivation and practices are simple and readily adoptable. It is a short-day plant (Rezaie *et al.*, 2005) and most varieties require high temperatures (Resi and Almodares, 2008) to make their best growth. The aim of agronomy in sweet sorghum is to increase productivity with focus on biofuel and improved feedstock supply as follows:

- Water and fertilizer (macro- and micro-nutrients) effects and their interaction on sugary stalk, grain and bagasse yield and quality.
- Effect of day length, temperature and their interaction, on stalk, grain and bagasse yield and quality.
- Crop rotation experiments to identify the most productive and sustainable cropping systems for different ecosystems.

Cultural Management

Ecological Requirements
Good surface drainage is preferred although sweet sorghum can withstand long waterlogged condition. Clay loam is preferred with soil acidity not lower than pH 6. However, it can tolerate a pH range of 5.0 to 8.0 (Smith and Fredriksen, 2000) and some

degree of salinity, alkalinity and poor drainage (Almodares and Hadi, 2009). It will also grow on heavy, deep cracking vertisols and light sandy soils (Smith and Frederiksen, 2000).

Land Preparation

Two rotavations at a depth of 25-30 cm are desirable to attain a fine and good soil tilth. This is necessary to have a uniform germination because sweet sorghum seeds are small as compared to corn.

Setting of Furrow

Two planting seasons are possible for sweet sorghum. During the wet season, furrows are set 100 cm apart while the dry season plantings are set 75 cm apart.

Planting

The recommended seeding rate is 5-8 kg per hectare to attain a population density of 130,000-150,000 plant/ ha. The seeds are drill-planted by hand or a planter can be used. The seed of sweet sorghum should be planted deep enough to give it moisture to germinate and allow its roots to grow down through moist soil into sub-soil moisture, ahead of the drying front (Almodares *et al.*, 2008b). Planting times usually start when the air temperature is above 12^0 C (Almodares and Mostafafi, 2006). Late planting reduces the length of the growing season, yield and carbohydrate content (Almodares *et al.*, 1994).

During the wet season planting, the furrows are set 10 cm deep. The seeds are drill planted at the bottom of the furrow and then spike tooth harrow is passed to cover the seeds to a depth of about 2-3 cm. For the dry season planting, the furrows should be made at least 15-20 cm deep to be able to make use of more residual soil moisture. The seeds are set at the bottom of the furrows but these are not covered anymore if the soil is dry. The impact of irrigation water running through a flexible hose which is directed at the side of the furrow will cover the seeds. In cases where the soil is moist, the technique used during the wet season planting is followed.

Thinning

For the 75 cm row spacing, 10-11 plants per meter are maintained, which is approximately 10 cm between plants with a population density of 130,000 plants/ ha. For the 100 cm spacing, 13 plants are maintained per meter of row or about 11 cm between plants with a population density of 150,000 plants/ ha. Thinning should be done before hilling-up or side dressing the second fertilizer dose which is 14-21 days after emergence.

Water Preparation

Sweet sorghum is remarkably drought-resistant and resists months of dry weather until rains resume. Supplemental irrigation is rarely needed but sweet sorghum needs moisture for uniform seed germination, hence, overhead irrigation is recommended at planting when moisture is insufficient for germination. Each fertilizer application should be followed by irrigation in case there is no rain.

Fertilization

A fertilizer rate of 80-60-60 (N-P-K) is generally recommended for a clay loam soil in both seasons. The basal fertilizer is 30-30-30 (N-P-K) or 215 kg of 14-14-14 (N-P-K) per hectare. This is 21-22g/linear meter of row in the 100 cm spacing and 16g/m in the 75 cm spacing. The fertilizer is drilled at the bottom of the furrow before planting. Balanced fertilization can increase yield (Rego et al., 2003). Nitrogen fertilizer and its application time (Almodares et al., 1996) promote sucrose content and growth rate in sweet sorghum (Tsialtas and Maslaris, 2005). Application of adequate amounts of K fertilizer increases yield responses better than increasing levels of N fertilizer alone (Pholsen and Sornsungnoen, 2004; Almodares and Mostafafi, 2006).

Pest and Disease Management

Pest

Shoot Fly

The shoot fly attacks soon after germination up to 30 days after planting (DAP). It is noted by dead hearts in seedlings and profuse tillering in affected plants. Carbofuran 3G is applied at 8-10 kg/ha during planting at the bottom of the furrow.

Stem Borer

Stem borer affects at a later stage up to maturity. Carbofuran 3G at 8-10 k/ha can be applied on leaf whorls (2-3 granules/whorl) to prevent stem borer tunneling.

Figure 1. Sweet sorghum plant with grains. Source: Bureau of Agricultural Research, Quezon City, Philippines.

Harvest Management

Stalks

Stalks may be harvested by hand, cut with a mower and picked up, or mechanically cut and squeezed in the field. The stalks are cut (similar to sugarcane) as close as possible to the ground leaving one node only. This node will be the sprouting point of the ratoon. These should be removed by hand after cutting the stalk down.

3. Sweet Sorghum Carbohydrate

The chief nonstructural carbohydrate in the grain of sweet sorghum is starch (Somani *et al.*, 1995), while sucrose is the main carbohydrate content in the stalk (Mamma *et al.*, 1996) that is the dominant form transported in the plant. Subramanian *et al.* (1994) have reported that cultivars with white or pale yellow seeds are most suitable for starch production. Both α- and β-amylases (Reisi and Almodares, 2008) are needed to hydrolyze starch and produce fermentable sugars (Beta *et al.*, 1995). The primary sugars present in grain of sweet sorghum are fructose, glucose, raffinose, sucrose and maltose (Somani *et al.*, 1995). In sorghum leaves, sucrose is translocated and transformed into starch during the development of grains (Smith and Frederiksen, 2000). The primary sugars present in sorghum stalk are sucrose, glucose, cellulose and hemicelluloses (Mamma *et al.*, 1996). Grain plus stem of sweet sorghum has been shown to yield more fermentable carbohydrates than other fuel crops (Murray *et al.*, 2008). In addition, the grain can be used for production of high fructose syrup (Hosseini *et al.*, 2003) and animal feed (Azarfa *et al.*, 1998). Therefore, sorghum is an excellent crop for biomass production.

In stems the extent of sucrose accumulation varies among cultivars (Almodares and Sepahi, 1996). Sorghum nonstructural carbohydrates contents are affected by temperature, time of day, maturity, cultivar, culm section, spacing and fertilization (Almodares and Hadi, 2009). Shading significantly reduces panicle and leaf laminae dry weights of sorghum (Kiniry *et al.*, 1992). Environmental conditions such as water quality and growth stage (Almodares *et al.*, 2008c) and maturities are important factors affecting carbohydrate contents (Almodares and Hadi, 2009). In the sweet sorghum, sucrose, glucose and fructose contents increase after anthesis (Almodares *et al.*, 2008c). In stems, nonstructural carbohydrates contents increase after preboot (Almodares *et al.*, 2008c) and reach a maximum level near post anthesis (Almodares *et al.*, 2008c).

4. Sweet Sorghum as a Bio-fuel Crop

One method to reduce air pollution is to produce oxygenated fuel for vehicles. MTBE (Methyl tert-butyl ether) is a member of a group of chemicals commonly known as fuel oxygenates (Fischer *et al.*, 2005). It is a fuel additive to raise the octane number. But it is very soluble in water and it is a possible human carcinogenic (Belpoggi *et al.*, 1995). Thereby, it should be substituted for other oxygenated substances to increase the octane number of the fuel. Presently, ethanol as an oxygenous biomass fuel is considered as a predominant alternative to MTBE for its biodegradable, low toxicity, persistence and regenerative characteristic (Cassada et al., 2000). The United States gasoline supply is an ethanol blend, and the importance of ethanol use is expected to increase as more health issues are related to air quality. Ethanol may be produced from many high energy crops such as sorghum, corn, wheat, barely, sugar cane, sugar beet, cassava and sweet potato (Drapcho et al., 2008). Like most biofuel crops, sweet sorghum has the potential to reduce carbon emissions. In addition, among the plants, sweet sorghum has the following special characteristics (Prasad et al., 2007):

- It is an efficient converter of solar energy, as it requires low inputs and yet, is a high carbohydrate producer.
- As a drought-tolerant crop with multiple uses.
- It has a concentration of sugar which varies between 16-22%.
- It can be cultivated in temperate, subtropical and tropical climates.
- All components of the plant have economic value - the grain from sweet sorghum can be used as food or feed, the leaves for forage, the stalk (along with the grain) for fuel, the fiber (cellulose) either as mulch or animal feed and with second generation technologies even for fuel (Gnansounou et al., 2005).
- Its bagasse, after sugar extraction, has a higher biological value than the bagasse from sugarcane, when used as feed for animals.
- Its growing period is shorter (3-5 months) than that of sugarcane (10-12 months), and the quantity of water required is one- third of sugarcane.
- In tropical irrigated areas sweet sorghum can be harvested twice each year (by ratooning) and its production can be completely mechanized.
- It has some tolerance to salinity
- It can produce large quantities of both readily fermentable carbohydrate, and fiber per unit land area.

Therefore, based on the above characteristics, it seems that sweet sorghum is the most suitable plant for biofuel production than other crops under hot and dry climatic conditions. In addition, possible use of bagasse as a by-product of sweet sorghum include: burning to provide heat energy, paper or fiber board manufacturing, silage for animal feed or fiber for ethanol production (Ban et al., 2008).

Sweet sorghum juice is assumed to be converted to ethanol at 85% theoretical, or 54.4 liters of ethanol per 100 kg fresh stalk yield. Potential ethanol yield from the fiber is more difficult to predict (Rains et al., 1993). The emerging enzymatic hydrolysis technology has not been proven on a commercial scale (Taherzadeh and Karimi, 2008). One ton of corn grain produces 387 L of 182 proof alcohol while the same amount of sorghum grain produces 372 L (Smith and Frederiksen, 2000). Sorghum is used extensively for alcohol production (Bulawayo et al., 1996; Smith and Frederiksen, 2000; Gnansounou et al., 2005), where it is significantly lower in price than corn or wheat (Smith and Frederiksen, 2000). The commercial technology required to ferment sweet sorghum biomass into alcohol has been reported in China (Gnansounou et al., 2005). One ton of sweet sorghum stalks has the potential to yield 74 L of 200- proof alcohol (Smith and Frederiksen, 2000).

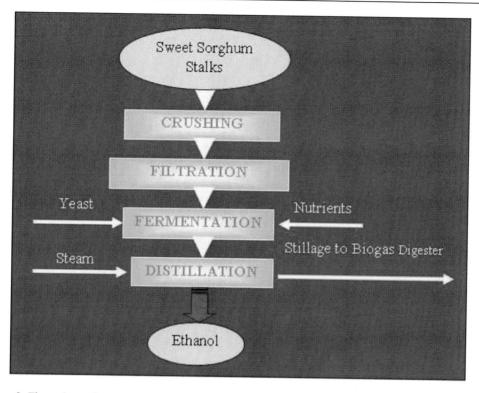

Figure 2. Flow-chart of ethanol production from sweet sorghum stalks. Source: Bureau of Agricultural Research, Quezon City, Philippines.

5. Processing Ethanol from Stalk

Juice Extraction

Juice is extracted from the stalk by a series of mills. The juice coming out of the milling section is first screened, sterilized by heating up to 100° C and then clarified (Quintero et al., 2008). The muddy juice is then sent to a rotary vacuum filter and the filtrate juice is sent to evaporation section for concentration (syrup to ethanol). The juice can also be directly sent to fermentation section (juice to ethanol). Depending on the scheme selected the juice can be concentrated using evaporators to attend various Brix. In the case of juice to ethanol (no syrup), it is advisable to partially increase the concentration of juice to 16–18° Brix. The syrup which needs storage for using during off season needs to concentrate to minimum 65° Brix (normally 85° Brix) (Almodares and Hadi, 2009).

Fermentation

Fermentation is a multidisciplinary process based on the chemistry, biochemistry and microbiology of the raw materials. Juice or syrup is converted into ethanol by microorganisms such as *Saccharomyces cerevisiae* (yeast) and *Zymomonas mobilis* (bacteria).

Sugar is converted to ethanol, carbon dioxide and yeast/bacterial biomass as well as much smaller quantities of minor end products such as glycerol, fusel oils, aldehydes and ketones (Laopaiboon et al., 2007; Jacques et al., 1999).

Distillation and Dehydration

In the distillation section, alcohol from fermented mash is concentrated up to 95% v/v. This is further concentrated to produce ethanol with 99.6% v/v (minimum) concentration. The treatment of vinasse generated in the distillation section can be done using the following option: concentration of part of vinasse to 20 to 25% solids followed by composting using press mud available and concentration of the rest of the vinasse to 55% solids and can be used as liquid fertilizer.

The schematic representation of ethanol fermentation from sweet sorghum stalk is given in Figure 2.

6. PROCESSING ETHANOL FROM GRAIN

The ethanol production processing from sweet sorghum grain and bagasse is similar to other starchy crops like corn and cassava (Quintero et al., 2008). Chemically starch is a polymer of glucose (Peterson, 1995). Yeast cannot use starch directly for ethanol production. Therefore, grain starch has to be completely broken down to glucose by a combination of two enzymes, viz., amylase and amyloglucosidase, before it is fermented by yeast to produce ethanol (Figures 2 and 3). After washing, crushing and milling the sweet sorghum grains, the starch is gelatinized, liquefied and saccharified using α-amylase and amyloglucosidase. Fermentation, distillation and dehydration processing of grain sorghum are similar to the sweet sorghum stalk. However, the by-products of grain are not similar to the stalk because DDGS (dried distillers grains with solubles) as a co-product of the ethanol production process from grain is a high nutrient valued feed which is used by the livestock industry. The step-wise processing of ethanol from sorghum grains are given below (Figure 4):

- Grain Crushing and Milling
- Cooking Process with enzymatic liquefaction
  - Gelatinization
  - Liquefaction and dextrinization (using dilute inorganic acid or thermostable α-amylase)
  - Saccharification (using amyloglucosidase).
- Fermentation
- Distillation

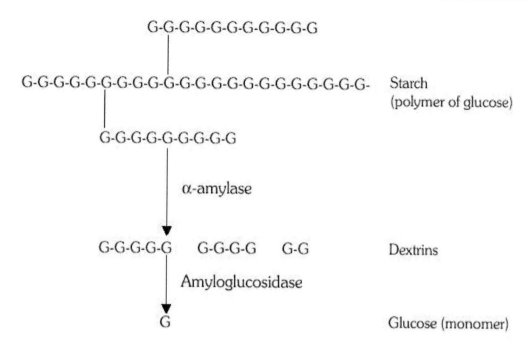

Figure 3. Enzymatic hydrolysis of starch to glucose.

Grain Crushing and Milling

Crushing and milling of grains are carried out by the following methods: (1) Dry milling - grinding of big solid particles into smaller ones (0.4 – 0.6mm), and (2) Wet milling - soaking in water to form lactic acid; separates germ from kernel. The dry milling has several advantages such as –

- less capital investment in plant and equipment
- fewer control loops and simpler processing
- shorter time from construction to operation
- minimal loss of starch

$$(C_6H_{10}O_5)_n + nH_2O \longrightarrow nC_6H_{12}O_6$$
$$\text{Starch} \qquad \text{Water} \qquad \text{Glucose}$$

$$C_6H_{12}O_6 \longrightarrow 2C_2H_5OH + 2CO_2$$
$$\text{Glucose} \qquad \text{Ethanol}$$

Figure 4. Chemical equation for conversion of starch to glucose and glucose to ethanol.

Gelatinization

The crushed and milled grains are made into mash by steam cooking, above starch gelatinization temperature (68-74^{0}C). The process is marked by melting of starch crystals, loss of birefringence, and starch solubilization. Granules absorb a large amount of water, swell to many times their original size, and open up enough for α-amylase to hydrolyze long chains into shorter dextrins (Ray and Ward, 2006).

Liquefaction/ Dextrinization

Breakdown of gelatinized starch into smaller fragments or dextrins is carried out by means of thermostable α- or β-amylase or dilute acid (Ray *et al.*, 2008). The action of α-amylase on gelatinized starch results in dramatic reduction of viscosity

- improves efficiency of the spiral or plate-and-frame heat exchangers used for cooling during saccharification or fermentation
- must be done immediately to prevent retrogradation or recrystallization of starch.

Fermentation and Distillation

The processes are the same as discussed for ethanol production from stalk.

7. PROCESSING ETHANOL FROM BAGASSE

The most promising future utilization of bagasse is cellulose-based ethanol production, while the residual solids (mainly lignin) can be burned to provide heat and power. Hydrolysis of the cellulose and hemicellulose fractions can be catalyzed by acids or cellulolytic enzymes. Enzymatic process needs a pretreatment step to increase the susceptibility of the cellulose, which can be degraded by cellulolytic enzymes to glucose (Gnansounou *et al.*, 2005).

Sweet sorghum residue was hydrolyzed with phosphoric acid under mild conditions. The liquid hydrolysate was fermented by *Pachysolen tannophilus*, and the hydrolysis residue was fermented by the simultaneous saccharification and fermentation (SSF) using *Saccharomyces cerevisiae* with cellulase (60 FPU/g dry materials) (Ban *et al.*, 2008). The results showed that the optimal reaction conditions were 120^0 C, 80 g/L, 80 min and 10%, respectively. Under these conditions, 0.3024 g reducing sugar/g dry bagasse was obtained. The liquid hydrolysate was then fermented by *P. tannophilus* with the highest ethanol concentration of 14.5 g/L. At a water-insoluble solid concentration of 5%, 5.4 g/L ethanol was obtained after 12 h of SSF. The total ethanol yield was 0.147 g/g dry bagasse.

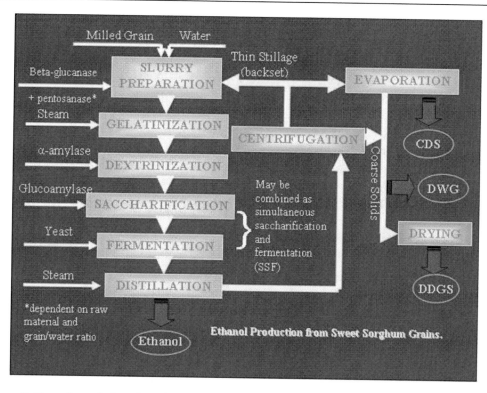

Figure 5. Flow-chart of ethanol production from sweet sorghum grains. Source: Bureau of Agricultural Research, Quezon City, Philippines.

In another study, sweet sorghum bagasse was steam-pretreated using various pretreatment conditions (temperatures and residence times). Efficiency of pretreatments was characterized by the degree of cellulose hydrolysis of the whole pretreated slurry and the separated fiber fraction. Two settings of the studied conditions (190^0 C, 10 min and 200^0 C, 5 min) were found to be efficient to reach conversion of 85–90%. Fiber separation followed by enzymatic digestion results in a readily fermentable hydrolysate (around 80–90% glucose to ethanol yields were achieved by *S. cerevisiae*). Moreover, applying a separation step before enzymatic hydrolysis, pentoses presented in the liquid fraction can be processed to several valuable products. This might be a basis for a lignocellulose bio-refinery (Sipos *et al.*, 2010).

8. PROCESSING BIO-DIESEL FROM SORGHUM

The unprecedented rise in diesel prices, along with increased consumption and vulnerability in the supply chain presents an ideal window of opportunities for bio-diesel. Several attempts have been made to use edible and non-edible oils in compression ignition for different utilities. But developing countries cannot afford to use edible oils as a power source; but non-edible sources such as sweet sorghum and others i.e., Jatropha, Pongamia, Neem, Kusum, are ideal and alternative bioresources.

A semi-solid fermentation process for the production of bio-diesel from sweet sorghum is introduced. The microorganism used is the oleaginous fungus *Mortierella isabellina*, which is able to efficiently transform sugar to storage lipid. The fungus consumed simultaneously sugars and nitrogen contained in sorghum and after nitrogen depletion the biomass growth was completed and oil accumulation began. Water content of 92% presented the highest oil efficiency of 11 g/100 g dry weight of substrate. The semi-solid process is shown to have certain advantages compared to liquid cultures or solid-state fermentation and gives oil of high quality (Economou *et al.*, 2010).

Microalga, *Chlorella protothecoides* could grow heterotrophically with glucose as the carbon source and accumulate a high proportion of lipids. The microalgal lipids are suitable for bio-diesel production. To further increase lipid yield and reduce bio-diesel cost, sweet sorghum juice was investigated as an alternative carbon source to glucose in the present study. When the initial reducing sugar concentration was 10 g/L in the culture medium, the dry cell yield and lipid content were 5.1 g/L and 52.5% using enzymatic hydrolyzates of sweet sorghum juice as the carbon source after 120 h-culture in flasks. The lipid yield was 35.7% higher than that using glucose. When 3.0 g/L yeast extract was added to the medium, the dry cell yield and lipid productivity was increased to 1.2 g/L/day and 586.8 mg/L/day. Bio-diesel produced from the lipid of *C. protothecoides* through acid catalyzed transesterification was analyzed by GC–MS, and the three most abundant components were oleic acid methyl ester, cetane acid methyl ester and linoleic acid methyl ester. The results indicated that sweet sorghum juice could effectively enhance algal lipid production, and its application may reduce the cost of algae-based bio-diesel (Gao *et al.*, 2010).

9. BIOTECHNOLOGY OF SORGHUM FERMENTATION

Solid Substrate Fermentation

Although ethanol is produced mostly by the process of submerged fermentation in various parts of the world, there is a tremendous scope for solid substrate fermentation. Solid substrate fermentation refers to the process whereby microbial growth and product fermentation occurs on the surface of the solid materials. This process occurs in the absence of "free" water, where the moisture is absorbed to the solid matrix. Solid substrate fermentation has several potential advantages: (1) less requirement of water (especially attractive in summer months when water is scarce), (2) smaller volumes of the fermentation mash, (3) less physical energy requirement, (4) smaller capital investment, fewer operating costs and lower space requirements, (5) reduced reactor volume, (6) easier product recovery, (7) less liquid water to be disposed of and hence less pollution problems (Ray *et al.*, 2008). There are some reports of ethanol production from sorghum using solid substrate fermentation.

Solid-substrate fermentation of chopped sweet sorghum particles to ethanol was studied in static flasks using an ethanol tolerant yeast (*S. cerevisiae*) strain. The influence of various process parameters, such as temperature, yeast cell concentration, and moisture content, on the rate and extent of ethanol fermentation was investigated. Optimal values of these parameters were found to be 35^0 C, 7×10^8 cells/g raw sorghum, and 70% moisture level, respectively (Kargi *et al.*, 1985a; Kiran Sree *et al.*, 1999). In a subsequent study, Kargi *et al.* (1985b), sugar compounds present in chopped solid-sweet sorghum particles were fermented to ethanol in a rotary drum fermentor using an ethanol tolerant yeast strain. The influence of rotational speed of the rotary drum fermentor on the rate of ethanol fermentation was investigated and compared with static flask experiments. The rate of ethanol formation decreased with increasing rotational speed. The maximum rate and extent of ethanol formation were ca. 3.1 g EtOH/L/h (based on expressed juice volume) and ca. 9.6 g EtOH/100 g mash, respectively, at 1 rpm rotational speed.

In solid substrate fermentation, ethanol productivity depends on a number of factors including particle size, moisture content, size of the inoculum, temperature and co-culturing two- or more microorganisms. Kiran Sree *et al.* (1999) studied ethanol production from sweet sorghum in solid substrate fermentation by co-culturing a thermotolerant *S. cerevisiae* and an amylolytic *Bacillus* sp. (VB9). More ethanol was produced in co-culture (5g EtOH/100g of substrate) than monoculture (3.5g EtOH/100g substrate) of the yeast alone. The other factors that influenced fermentation were temperature, 37^0 C, moisture, 70%, substrate particle size, 60-100 μ and inoculum size, 10%.

The sugars in unsterilized sweet sorghum solids fermented readily during anaerobic solid-state fermentation. Products, based on 100 g hexose, were: 36.7 g volatiles (mass loss), 32.8 g ethanol, 2.8 g glycerol, 12.7 g lactic acid, 2.0 g acetic acid, 4.8 g dextrans, and 4.2 g mannitol, presumably from fermentation with naturally occurring heterolactic bacteria, such as *Leuconostoc dextranicum*, as well as wild yeasts. Inoculation of unsterilized sorghum solids with 0.007% to 0.20% *Saccharomyces cerevisiae* increased fermentation rate, gas evolution, and yields of ethanol, glycerol, and acetic acid, while other yields decreased. With 0.2% yeast, yields were 40.9 g ethanol, 4.8 g glycerol, and 45.5 g gases (Bryan, 1990).

Experiment of solid-state fermentation (SSF) for ethanol using dry sweet sorghum stalk particles with active dry yeast were carried out to clarify the effect of different variables, including temperature, particle size, yeast inoculation rate, and water content on yeast growth, CO_2 and ethanol formation, and sugar use (Shen and Liu, 2009). The results indicated that the effect of temperature and particle on yeast growth and ethanol yield were significant. In addition, the yeast inoculation rate and water content were greatly related to yeast growth, even through they had less significant effect on ethanol yield. The suitable ranges of temperature and particle size for yeast growth were 25-30^0 C and 1.6-2.5 mm, respectively, in which higher values of $Y_{cell/sugar}$ of 0.2681 and 0.3538 mg/mg were obtained. The suitable ranges for ethanol production were 35-40^0 C and 0.9-1.6 mm, in which higher values of $Y_{ethanol/sugar}$ of 0.2404 and 0.2702 mg/mg could be obtained. The adequate yeast inoculation rate should be 0.250, and higher values of $Y_{ethanol/sugar}$ and $Y_{cell/sugar}$ could be up to 0.2486 and 0.3017 mg/mg, respectively. In addition, the suitable water content for ethanol production and yeast growth was 76.47%.

Simultaneous Saccharification and Fermentation (SSF)

The most important process development made for enzymatic hydrolysis of various biopolymers (i.e., starch, cellulose and hemi-cellulose) - containing crops and biomass is the introduction of simultaneous saccharification and fermentation (SSF) process (Ray and Naskar, 2008). This process employs thermotolerant yeast strains to reduce the number of reactors involved by eliminating the separate (saccharification) reactor and more importantly, avoiding the problem of product elimination associated with enzymes such as build up of cellobiase and amyloglucosidase which shut down cellulose and starch degradation, respectively (Chandel *et al.*, 2007). However, in SSF both saccharifying enzyme and fermenting microorganisms are applied simultaneously. As the conversion of starch, cellulose and hemi-cellulose into sugars is processed by respective microbes or their enzymes, the fermentative organisms convert them into ethanol.

A few fungal species such as *Neurospora cra*ssa and *Fusarium oxysporium* were reported in the 1980s to have the ability to ferment cellulose directly to ethanol by SSF (Deshpande *et al.*, 1986). The SSF of sweet sorghum carbohydrates to ethanol by *Fusarium oxysporum* F3 alone or in mixed culture with *S. cerevisiae* 2541 or *Z. mobilis* CP4 in a fed-batch fermentation process was studied. While SSF was adequately carried out by the first microorganism the process achieved its maximum value by the mixed culture of the fungus and yeast. Under optimum conditions, ethanol yields and concentrations as high as 29.7 g of ethanol per 100 g of dry sorghum stalk and 7.5% (w/v) respectively were obtained. These values together with the high yield of sorghum crop in Greece make this process promising and worthy of further investigation for the production of fuel bio-ethanol (Christakopoulos *et al.*, 1994; Lezinou *et al.*, 1994). In another study, sweet sorghum carbohydrates were simultaneously saccharified and fermented to ethanol by a mixed culture of *F. oxysporum* and *S. cerevisiae* in a bioreactor. *Fusarium oxysporum* was grown aerobically for the production of the enzymes necessary for the saccharification of sorghum cellulose and hemicellulose. *Saccharomyces cerevisiae*, together with *F. oxysporum*, converted the soluble sugars to ethanol. Three batches of sorghum were used, harvested at different periods of the year. The optimum yield of bioconversion and ethanol concentration was 4 g ethanol/100 g of fresh sorghum and 9% (w/v), respectively, depending on the composition of sorghum stalks. In all experiments, the ethanol yield exceeded the theoretical, based on soluble sugars, by 1% due to bioconversion of polysaccharides to ethanol (Mamma *et al.*, 1996).

Cell Immobilization

In recent years, immobilization of enzymes or microbial cells or co-immobilization of both for the simultaneous starch saccharification and fermentation process has attained scientific and technical interest for alcoholic fermentation. By concentrating an active cell biomass or enzyme in a bioreactor, the efficiency of bioconversion increases, as does the reactor productivity which, in turn, results in the reduction of the reactor size for a given production rate (Nunez and Lema, 1987). Immobilization can be carried out in different ways; adsorption and entrapment in matrices are the methods most commonly used (Kim and Rhee, 1993; Yu *et al.*, 1996).

Shen *et al.* (2009) studied the main factors, including temperature, agitation rate, pH, and particles stuffing rate for ethanol fermentation from stalk juice of sweet sorghum using immobilized *S. cerevisiae* in shaking flasks. The results showed that a temperature of 34^0 C, agitation rate of 150-200 revolution/min, pH of 4.5, and particles stuffing rate of 25% should be selected as the suitable condition. The results of verification experiments at the selected condition demonstrated that the ethanol yield, the CO_2 weight loss rate, and fermentation time were 95.15%, 0.508 g/h, and 6 h in the shaking flasks, respectively. Meanwhile, the ethanol yield and the fermentation time were 96.72% and 10 h in the 5 L bioreactor. As a result, it could be concluded that the determined condition was suitable and reasonable for ethanol fermentation from by immobilized *S. cerevisiae*.

Liu *et al.* (2008) conducted bio-ethanol fermentation experiments in shaking flasks and 10 L fluidized bed bioreactor with stalk juice of Yuantian NO. 1 sweet sorghum cultivar when immobilized yeast was applied. The experimental results in the shaking flasks showed that the order of influence on improving ethanol yield was $(NH_4)_2SO_4 > MgSO_4 > K_2HPO_4$, and the optimum inorganic salts supplement dose was determined as follows: K_2HPO_4 0%, $(NH_4)_2SO_4$ 0.2%, $MgSO_4$ 0.05%. When the optimum inorganic salts supplement dose was used in fermentation in a 10 L fluidized bed reactor, the fermentation time and ethanol content were 5 h and 6.2% (v/v), respectively, and ethanol yield was 91.6%, which was increased by 9.7% from the control. In addition, the result showed that the fermentation time was about 6-8 times shorter in fluidized bed bioreactor with immobilized yeast than that of conventional fermentation technology. As a result, it can be concluded that the determined optimum inorganic salts supplement dose could be used as a guide for commercial ethanol production. The fluidized bed bioreactor with immobilized yeast technology has a great potential for ethanol fermentation of stalk juice of sweet sorghum.

Use of Adjuncts

Ethanol production from sweet sorghum juice by *Saccharomyces cerevisiae* NP01 was investigated under very high gravity (VHG) fermentation and various carbon adjuncts and nitrogen sources (Laopaiboon *et al.*, 2009). When sucrose was used as an adjunct, the sweet sorghum juice containing total sugar of 280 g/L, 3 g yeast extract/L and 5 g peptone/L gave the maximum ethanol production efficiency with concentration, productivity and yield of 120.7 g/L, 2.1 g/L/h and 0.5 g/g, respectively. When sugarcane molasses was used as an adjunct, the juice under the same conditions gave the maximum ethanol concentration, productivity and yield with the values of 109.3 g/L, 1.5 g/L/h and 0.5 g/g, respectively. In addition, ammonium sulphate was not suitable for use as a N supplement in the sweet sorghum juice for ethanol production since it caused the reduction in ethanol concentration and yield for approximately 14% when compared to those of the un-supplemented juices.

10. ECONOMICS OF SWEET SORGHUM BIO-FUEL

Guo *et al.* (2010) studied the bio-ethanol agro-industrial production system in sweet sorghum in China. The system consists of the following processes: sweet sorghum cultivation, crude ethanol production, ethanol refining and by-product utilization. The plant capacities of crude ethanol and pure ethanol, in different fractions of useful land, are optimized. Assuming a minimum cost of investment, transport, operation and so on, the optimum capacity of the pure ethanol factory is 50,000 tons/year. Moreover, this bio-ethanol system, which requires *ca.* 13,300 ha (hectares) of non-cultivated land to supply the raw materials, can provide 26,000 jobs for rural workers. The income from the sale of the crops is approximately 71 million US$ and the ethanol production income is approximately 13.84

million US$. The potential savings in CO_2 emissions are *ca.* 423,000 tons/year and clear economic, social and environmental benefits can be realized (Gnansounou *et al.*, 2005).

Under socio-economic conditions prevailing in India, the cost of ethanol production per liter from sweet sorghum juice is comparable with that of sugarcane molasses and significantly less than with maize and cassava as feedstock (Reddy *et al.*, 2007). Sorghum can potentially give a good yield of alcohol of about 380 to 390 L absolute alcohol/ton of grain provided the process is optimized. Total installed capacity for potable alcohol production using both grain and molasses is 110 million L/year (Sheorain *et al.*, 2000). If the average yield of alcohol production is 390 L/ton of grain, approximately 0.28 million tons of sorghum will be required every year if all the potable alcohol was to be produced from sorghum. This will provide an uptake of 10–15% of available damaged grain annually and help create a market for damaged grain thereby considerably helping sorghum producers. Similar opinions have been advocated for countries like Iran (Almodares and Hadi, 2009) and Hungary (Sipos *et al.*, 2010). Hybrid seed technology can create a major advance in economic competitiveness, as well as in productivity of both grain and bio-ethanol. Like sugarcane, sweet sorghum bio-ethanol system also yields a highly positive net energy balance [Energy output/fossil energy input estimated at approximately 8.0], roughly four times higher than for maize grain or cassava as feedstock.

CONCLUSIONS AND FUTURE PERSPECTIVES

Ethanol demand is increasing drastically in the present time due to its blending in automotive fuels, which is desirable for getting clean exhaust and fuel sufficiency. The higher cost of cultivation of sugarcane/beets, highly sensitive molasses rates, and ultimately instabilities in the price of ethanol have created grounds to search for an alternative source for ethanol production. Sweet sorghum has shown potential as a raw material for fuel-grade ethanol production due to its rapid growth rate and early maturity, greater water use efficiency, limited fertilizer requirement, high total value, and wide adoptability. Ethanol-producing companies, research institutions, and governments can coordinate with farmers to strategically develop value-added utilization of sweet sorghum. Fuel-grade ethanol production from sweet sorghum syrup can significantly reduce dependence on petroleum (gasoline) to some extent and also minimize the environmental threat caused by fossil fuels. Sweet sorghum can also prove to be an ideal substrate for bio-diesel production.

ACKNOWLEDGMENTS

The permission to print the photographs 1, 2 and 5 by The Technology Commercialization Unit, Bureau of Agricultural Research, RDMIC Bldg. Visayas Avenue cor. Elliptical Road, Diliman, Quezon City Philippines 1104, is sincerely acknowledged.

REFERENCES

Almodares, A., Aghamiri, A. and Sepahi, A. (1996). Effects of the amount and time of nitrogen fertilization on carbohydrate contents of three sweet sorghum cultivars. *Ann. Plant Physiol.* 10: 56-60.

Almodares, A. and Hadi, M.R. (2009). Production of bioethanol from sweet sorghum: a review. *Afr. J. Agric. Res.* 4: 772-780.

Almodares, A., Hadi, M.R. and Ahmadpour, H. (2008a). Sorghum stem yield and soluble carbohydrates under phenological stages and salinity levels. *Afr. J. Biotechnol.* 7: 4051-4055.

Almodares, A., Hadi, M.R. and Dosti, B. (2007). Effects of salt stress on germination percentage and seedling growth in sweet sorghum cultivars. *J. Biol. Sci.* 7: 1492-1495.

Almodares, A. and Mostafafi, D.S.M. (2006). Effects of planting date and time of nitrogen application on yield and sugar content of sweet sorghum. *J. Environ. Biol.* 27: 601-605.

Almodares, A. and Sepahi, A. (1996). Comparison among sweet sorghum cultivars, lines and hybrids for sugar production. *Ann. Plant Physiol.* 10: 50-55.

Almodares, A., Sepahi, A., Dalilitajary, H. and Gavami, R. (1994). Effect of phenological stages on biomass and carbohydrate contents of sweet sorghum cultivars. *Ann. Plant Physiol.* 8: 42-48.

Almodares, A., Taheri, R. and Adeli, S. (2008b). Stalk yield and carbohydrate composition of sweet sorghum [*Sorghum bicolor* (L.) Moench] cultivars and lines at different growth stages. *J. Malaysian Appl. Biol.* 37: 31-36.

Almodares, A., Taheri, R. and Safavi, V. (Eds.) (2008c). *Sorghum, Botany, Agronomy and Biotechnology,* Jahad Daneshgahi of University of Isfafan, Isfahan, Iran.

Azarfa, A., Ghorbani, G., Asadi, Y. and Almodares, A. (1998). Dry matter, organic matter and starch degradability in barley and three low, medium and high tannin sorghum varieties. *Agr. Sci.* 8: 56-60.

Ban, J., Yu, J., Zhang, Xu and Tan, T. (2008). Ethanol production from sweet sorghum resudial. *Frontiers Chemical Eng. China* 2: 452-455.

Behera, S., Kar, S., Mohanty, R.C. and Ray, R.C. (2010a). Comparative study of bio-ethanol from mahula (*Madhuca latifolia* L.) flowers by *Saccharomyces cerevisiae* cells immobilized in agar agar and Ca-alginate matrices. *Appl. Energy.* 87: 96-100.

Behera, S., Mohanty, R.C. and Ray, R.C. (2010b). Comparative study of bio-ethanol production from mahula (*Madhuca latifolia* L.) flowers by *Saccharomyces cerevisiae* and *Zymomonas mobilis*. *Appl. Energy* 87: 2352-2355.

Belpoggi, F., Soffritti, M. and Maltoni, C (1995). Methyl tertiary-butyl ether (MTBE) - A gasoline additive - causes testicular and lymphohaematopoetic cancers in rats. *Toxicol. Ind. Health* 11: 119-149.

Beta, T., Rooney, L. and Waniska, R. (1995). Malting characteristics of sorghum cultivars. *Cereal Chem.* 72: 533-538.

Bryan, W.L. (1990). Solid state fermentation of sugars in sweet sorghum. *Enzyme Microb. Technol.* 12: 437-442.

Bulawayo, B., Bvochora, J. M., Muzondo, M. I. and Zvauya, R. (1996). Ethanol production by fermentation of sweet sorghum juice using various yeast strains. *World J. Microbiol. Biotechnol.* 12: 357-360.

Cassada, D.A., Zhang, Y., Snow, D.D. and Spalding, R. F. (2000). Trace analysis of ethanol, MTBE, and related oxygenate compounds in water using solid-phase microextraction and gas chromatography/mass spectrometry. *Analyt. Chem.* 72: 4654-4658.

Chandel, A.K., Chan, E.S., Rudravaram, R., Narasu, M.L., Rao, L.V. and Ravindra, P. (2007). Economics and environmental impact of bioethanol production technologies: an appraisal. *Biotechnol. Mol. Biol. Rev.* 2(1):14- 32.

Christakopoulos, P., kekos, D. and Macris, B. J. (1994). Simultaneous saccharification and fermentation of sweet sorghum carbohydrates to ethanol in a fed-batch process. *Biotechnol. Lett.* 16: 983- 988.

Deshpande, V., Sulbha, K., Mishra, C. and Rao, M. (1986). Direct conversion of cellulose/hemicellulose to ethanol by *Neurospora crassa*. *Enzyme Microb. Technol.* 8: 149- 152.

Drapcho, C. M., Nhuan, N. P. and Walker, T.H. (2008). *Biofuels Engineering Process Technology,* The McGraw-Hill companies, Inc, U.S.A..

Economou, C., Aggelis, M.A., Paviou, G. and Vayenas, D.V. (2010). Semi-solid state fermentation of sweet sorghum for the biotechnological production of single cell. *Biores. Technol.* 101: 1385- 1388.

Gao, C., Zhai, Y., Ding, Yi and Wu, Q. (2010). Application of sweet sorghum for biodiesel production by heterotrophic microalga *Chlorella protothecoides*. *Appl. Energy* 87: 756-761.

Gnansounou, E., Dauriat, A. and Wyman, C. E. (2005). Refining sweet sorghum to ethanol and sugar: economic trade-offs in the context of North China. *Biores. Technol.* 96: 985-1002.

Guo, Y., Hu, S., Li, Y., Chen, D., Zhu, B. and Smith, K.M. (2010). Optimization and analysis of a bioethanol agro-industrial system from sweet sorghum. *Renewable Energy* doi:10.1016/j.renenl.2010.04.024.

Fischer, A., Oehm, C., Selle, M. and Werner, P. (2005). Biotic and abiotic transformations of methyl tertiary butyl ether (MTBE). *Environ. Sci. Pollut. Res. Int.* 12: 381-386.

Hosseini, H., Almodares, A. and Miroliaei, M. (2003). Production of fructose from sorghum grain. In: *Proceeding of the 11th Iranian Biology Conference* (H. Hosseini, A. Almodares and M. Miroliaei, Eds.), Urmia, Iran.

Jacques, K., Lyons, T. P. and Kelsall, D. R. (Eds.) (1999). *The Alcohol Textbook*. Nottingham University Press,UK, 3rd Edition, 388pp.

Kargi, F. and Curme, J.A. (1985b). Solid state fermentation of sweet sorghum to ethanol in a rotary-drum fermentor. *Biotechnol. Bioeng.* 27: 1122- 1125.

Kargi, F., Curme, J.A. and Sheehan, J. (1985a). Solid state fermentation of sweet sorghum to ethanol. *Biotechnol. Bioeng.* 27: 34- 40.

Kim, C.H. and Rhee, S.K. (1993). Process development for simultaneous starch saccharification and ethanol fermentation by *Zymomonas mobilis*. *Process Biochem.* 28 : 331-339.

Kiniry, J.R., Tischler, C. R., Rosenthal, W. D. and Gerik, T. J. (1992). Nonstructural carbohydrate utilization by sorghum and maize shaded during grain growth. *Crop Sci.* 32: 131-137.

Kiran Sree, N., Sridhar, M., Rao, L. V. and Pandey, A. (1999). Ethanol production in solid substrate fermentation using thermotolerant yeast. *Process Biochem.* 34: 115- 119.

Laopaiboon, L., Thanonkeo, P., Jaisil, P. and Laopaiboon, P. (2007). Ethanol production from sweet sorghum juice in batch and fed-batch fermentations by *Saccharomyces cerevisiae*. *World J. Microbiol. Biotechnol.*, 23: 1497-1501.

Laopaiboon, L., Nuanpeng, S., Srinophakun, P., Klanrit, P. and Laopaiboon, P. (2009). Ethanol production from sweet sorghum juice using very high gravity technology: effects of carbon and nitrogen supplementations. *Biores. Technol.* 100: 4176- 4182.

Lezinou, V., Christakopoulos, P., Kekos, D. and Macris, B.J. (1994). Simultaneous sacharification and fermentation of sweet sorghum carbohydrates to ethanol in a fed-batch process. *Biotechnol. Lett.* 16: 983- 988.

Liu, R., Li, J. and Shen, F. (2008). Refining bioethanol from stalk juice of sweet sorghum by immobilized yeast fermentation. *Renewable Energy* 33: 130 -135.

Mamma, D., Koullas, D., Fountoukidis, G., Kekos, D., Macris, B.J. and Koukios, E. (1996). Bioethanol from sweet sorghum: Simultaneous saccharification and fermentation of carbohydrates by a mixed microbial culture. *Process Biochem.* 31: 377- 381.

Murray, S.C., Sharma, A., Rooney, W.L., Klein, P.E., Mullet, J.E., Mitchell, S.E. and Kresovich, S. (2008). genetic improvement of sorghum as a biofuel feedstock: I. QTL for stem sugar and grain nonstructural carbohydrates. *Crop Sci.* 48: 2165-2179.

Nadir, N., Mel, M., Karim, M.I.A. and Yunus, R.M. (2009). Comparison of sweet sorghum and cassava for ethanol production by using *Saccharomyces cerevisiae*. *J. Applied Sci.* 9: 3068-3073.

Nunez, M.J. and Lema, J.M. (1987). Cell immobilization: Application to alcoholic production. *Enzyme Microb. Technol.* 9: 642-650.

Peterson, A. (1995). Production of fermentable extracts from cereals and fruits. In: *Fermented Beverage Production* (A.G.H. Lea and J. R. Piggot, Eds.), Blackie Academic and Professional., London, UK, pp. 1-31.

Pholsen, S. and Sornsungnoen, N. (2004). Effects of nitrogen and potassium rates and planting distances on growth, yield and fodder quality of forage sorghum (*Sorghum bicolor* L. Moench). *Pakistan J. Biol. Sci.* 7: 1793-1800.

Prasad, S., Singh, A., Jain, N. and Joshi, H.C. (2007). Ethanol production from sweet sorghum syrup for utilization as automotive fuel in India. *Energy Fuel* 21: 2415- 2420.

Quintero, J. A., Montoya, M. I., Sanchez, O.J., Giraldo, O. H. and Cardona, C. A. (2008). Fuel ethanol production from sugarcane and corn: Comparative analysis for a Colombian case. *Energy* 33: 385-399.

Rains, G. C., Cundiff, J. S. and Welbaum, G. E. (1993). Sweet sorghum for a piedmont ethanol industry. In: *New Crops* (J. Janick and J.E. Simon, Eds.). Wiley, New York, U.S.A., pp. 394–399.

Ray, R.C. and Naskar, S.K. (2008). Bio-ethanol production from sweet potato (*Ipomoea batatas* L.) by enzymatic liquefaction and simultaneous saccharification and fermentation. *Dyn. Biotechnol. Process Biochem.Mol. Biol.* 2: 47- 49.

Ray, R. C., Mohapatra, S., Panda, S. and Kar, S. (2008). Solid substrate fermentation of cassava fibrous residue for production of α- amylase, lactic acid and ethanol. *J. Environ. Biol.* 29 (1/2):111-115.

Ray, R.C. and Ward, O.P. (2006). Post- harvest microbial biotechnology of topical root and tuber crops. In: *Microbial Biotechnology in Horticulture, Volume 1* (Ramesh C. Ray and O.P. Ward, Eds.), Science Publishers, New Hampshire, U.S.A., pp. 345- 396.

Reddy, B.V.S., Ramesh, S., Reddy, P.S., Ramaiah, B., Salimath, P.M. and Kachapur, R. (2005). Sweet Sorghum–A Potential Alternate Raw Material for Bio-ethanol and Bio-energy. *Int. Sorghum Millets Newslett.* 46: 79–86.

Reddy, B.V.S., Ashok Kumar, A. and reddy, P.S. (2007). Sweet sorghum: A dryland-adapted, pro-poor bioethanol feedstock yielding both grain and fuel. Abstract of the paper presented at *USDA Biofuel Conference*, 20-22 August 2007, Minneeapolis, Minnesota, U.S.A..

Rego, T.J., Rao, V.N., Seeling, B., Pardhasaradhi, G. and Rao, J.V.D.K. (2003). Nutrient balances a guide to improving sorghum and ground based dry land cropping systems in semi-arid topical India. *Field Crop Res.* 81: 53-68.

Reisi, F. and Almodares, A. (2008). The effect of planting date on amylose content in sorghum and corn. In: *Proceeding of the 3rd International biology conference* (F. Reisi and A. Almodares, Eds.), Tehran, Iran.

Rezaie, A., Almodares, A., Amini, M. and Khanghly, S. (2005). The effect of milo photoperiodic to red light In: *Proceeding of the 1st National Congress on Forage Grass* (A. Rezaie, A. Almodares, M. Amini and S. Khanghly, Eds.), University of Tehran Agriculture and Natural Resources, Karaj, Iran.

Shen, F. and Liu, R. (2009). Research on solid state ethanol fermentation using dry sweet sorghum stalk particles with active dry yeast. *Energy Fuels* 23: 519- 525.

Shen, F., Liu, R. and Wang, T. (2009). Effects of temperature, pH, agitation and particle stuffing on fermentation of sorghum stalk juice to ethanol. *Energy Sources. Part A. Recovery, Utilization and Environmental Effects* 31: 646- 656.

Sheorain, V., Banka, R. and Chavan, M. (2000). Ethanol production from sorghum. In: *Technical and Institutional Options for Sorghum Grain Mold Management: Proceedings of an International Consultation*, (A. Chandrashekar, R. Bandyopadhyay and A. J. Hall, Eds.), 18-19 May 2000, International Crops Research Institute for the Semi-Arid Tropics (ICRISAT), Patancheru, India, pp. 228-239.

Sipos, B., Réczey, J., Somorai, Z., Kádár, Z., Dienes, D. and Réczey, K. (2010). Sweet sorghum as feedstock for ethanol production:enzymatic hydrolysis of steam-pretreated bagasse. *Appl. Biochem. Biotechnol.* DOI 10.1007/s12010-008-8423-9.

Smith, C.W. and Frederiksen, R.A. (Eds.) (2000). Sorghum: Origin, history, technology, and production, John Wiley and Sons, New York.

Somani, R.B., Almodares, A. and Shirvani, M. (1995). Preliminary studies on sweetener production from sorghum grains. *Ann. Plant Physiol.* 9: 146-148.

Subramanian, V., Hoseney, R.C. and Cox, P.B. (1994). Factors affecting the color and appearance of sorghum starch. *Cereal Chem.* 71: 275-278.

Taherzadeh, M. J. and Karimi, K. (2008). Bioethanol: Market and Production Processes. In: Biofuels Refining and Performance (A. Nag, Ed.). The Mc Graw- Hill, U.S.A., pp. 1621-1655.

Tesso, T.T., Claflin, L.E. and Tuinstra, M.R. (2005). Analysis of stalk rot resistance and genetic diversity among drought tolerant sorghum genotypes. *Crop Sci.* 45: 645-652.

Tsialtas, J.T. and Maslaris, N. (2005). Effect of N fertilization rate on sugar yield and non-sugar impurities of sugar beets *(Beta vulgaris)* grown under Mediterranean conditions. *J. Agron. Crop Sci.* 191: 330-339.

Yu, B., Zhang, F., Zheng, Y. and Wang, P. (1996). Alcohol fermentation from the mash of dried sweet potato with its dregs using immobilized yeast. *Process Biochem* 31 (1) : 1-6.

Chapter 3

SWEET SORGHUM AS AN ENERGY CROP

Yanna Liang
Department of Civil and Environmental Engineering, Southern Illinois University Carbondale, Carbondale, Illinois, U.S.A.

ABSTRACT

As a potential energy crop, sweet sorghum has attracted considerable attention during recent years. Three kinds of biofuels, ethanol, lipid, and hydrogen can be produced from sweet sorghum through different processes. Whole stalk of sweet sorghum, juice, and bagasse have been explored to generate these biofuels to certain extent. Based on related publications, the best technologies for different biofuel production processes are identified for further investigation. Logistic issues, such as sorghum harvesting and storage are also discussed in this chapter. Energy and economic analyses are described without too many details considering the regionally specific features. In summary, every part of sweet sorghum can be utilized for biofuel production. The cost-benefit and life cycle analysis, however, needs to be investigated thoroughly to ensure that plantation of sweet sorghum for bioenergy purposes is sustainable, economical, and scalable.

1. INTRODUCTION

Sorghum which appears similar to maize, is a coarse tropical grass and is mostly self pollinating. The major sorghum species used world wide is *Sorghum bicolor*, which covers all of the grain sorghum, sweet sorghum, broom corns, and many forage sorghums [1]. Compared to switchgrass and wheat, sorghum is the second most important feed grain grown in the U.S. in terms of acreage [2]. Sweet sorghum is a crop close to sugarcane in terms of its sugar accumulation in the stems [3]. Like other sorghum types, sweet sorghum probably originated from East Africa and spread to other African regions, Southern Asia, Europe, Australia, and the U.S. [4]. Currently, sweet sorghum is planted in 99 countries around the world in 44 million ha, mainly in poor and semi-arid areas [5]. Sweet sorghum has many unique characteristics: 1) wide adaptabililty. Despite being a native to tropics, sweet sorghum has well adapted to temperate climate [4]. Sweet sorghum varieties are even adapted for production as far north as Wisconsin in the U.S.A. [6]; 2) high photosynthetic efficiency and high sugar yield [7, 8]. As a C4 plant, it returns 4-5 kJ of food and fiber product for each kJ of energy used [1]; 3) resistance to abiotic stresses like drought, water logging, salinity, and alkalinity [9, 10]; 4) less fertilizer requirement than corn [3]; and 5) rapid growth rate and the ability to reach maturity in 3-5 months [11]. Depending on the varieties and growing conditions, sweet sorghum can be an annual or short perennial crop [4]. Seeds are generally sown in Spring when the soil temperature is 15-18°C. The plant can grow to a height of 120-400 cm.

During recent years, sweet sorghum has drawn intensive attentions world-wide due to its potential as an energy crop. Both the stalk juice and the left bagasse can be processed for various liquid transportation fuels. In a recent request for application issued by USDA (http://www.nifa.usda.gov/funding/rfas/pdfs/10_afri_bioenergy.pdf.), sweet sorghum is listed as one of the energy feedstocks suitable for the south central and southeast U.S.A. Regarding the southeastern region, sweet sorghum requires the least inputs of land, water, and nitrogen fertilization when compared to those for corn and sugarcane [12]. Globally speaking, sweet sorghum has been actively studied for fuel production in Thailand [13], India [3, 10, 14], Egypt [15], Japan [16], Zambia [17], Italy [18], Greece [19, 20], Hungary [21], Taiwan where it is strongly recommended as a key alcohol crop [22], and China where it has the greatest potential as an energy crop among all the non-food feedstocks [23-25]. In China, sweet sorghum is also identified as the most sustainable when evaluated on nine sustainability indicators which focus on resource use efficiency, soil quality, net energy production, and greenhouse gas emissions while disregarding socio-economic or biodiversity aspects and land use change [26]. In this chapter, the use of sweet sorghum for producing ethanol, lipids, and hydrogen will be discussed in detail. Future research direction in this field is provided as well.

2. BIOFUEL PRODUCTION

2.1. Ethanol Production from Sorghum

Different parts of sweet sorghum have been used for ethanol production. Extracted juice which has been traditionally used for producing edible syrup in the U.S.A. for more than 100 years [27] has been the main material for this purpose. But nowadays, responding to the second generation cellulosic ethanol, sorghum bagasse has also been investigated for producing ethanol. Few studies have even evaluated ethanol yield from unsqueezed stalk. The following section focuses on discussing ethanol production from the juice, bagasse, and the whole stalk.

2.1.1. Producing Ethanol from Sorghum Juice

The sugar yield of sweet sorghum is highly attractive. Typically, the fresh stem yield of sweet sorghum is more than 45 t/ha per year, and the total sugar obtained is 5.1-10.5 t/ha considering the sugar content ranging from 11% to 23% [4, 28]. Sorghum juice is mainly composed of three kinds of sugars: fructose, glucose, and sucrose. Depending on the sorghum variety, location, and harvest time, the composition of the juice and the total sugar concentration can be very different. As summarized by Liang, KN Morris so far has the highest sugar content as 24.3% [29].

Considering the high sugar content in the juice, it is intuitive to attempt to convert the sugars to ethanol through yeast fermentation, a process broadly used for ethanol production from sugar cane. This approach has been explored extensively [10, 13, 30-35] as detailed in Table 1. As indicated by this Table, producing ethanol from sorghum juice is a readily available technique. No matter batch or fed-batch, agitated or static, or different strains of *Saccharomyces cerevisiae* used, most studies have achieved an ethanol yield higher than 93%. One study by Laopaiboon reported a 100% ethanol yield when a very high gravity (VHG) technology was used [33]. VHG is referred to a content of dissolved solids higher than 270 g/l. This technique intends to decrease capital costs, energy costs, and the risk of bacterial contamination while increasing ethanol concentration in the fermenting broth and the rate of fermentation. In this investigation, the sorghum juice was supplemented with sucrose to bring the final sugar concentration to 24%, 28%, and 32% from the original 18%. Different nitrogen sources, such as yeast extract and ammonium sulfate were evaluated in terms of ethanol production. The highest ethanol concentration, 120.68 g/l was observed when the dissolved solid content was 28% with yeast extract as the nitrogen source. However, not all sugars were consumed after 60 h fermentation. Under this optimal condition, the final total sugar concentration was 47.7 g/l. To use this VHG technique, other materials, such as starch or dextrins may also be used to raise the sugar concentration, but additional steps for hydrolyzing the complex carbohydrates may be needed. The best application of this VHG technology may be when sorghum syrup is used for fermentation. On one hand, evaporating sorghum juice to a high sugar concentration is good for preservation. On the other hand, syrup fermentation for producing ethanol can be operated year round at high efficiency using this VHG approach.

Table 1. Comparison of ethanol production from different sorghum juice samples

Sugar conc. (%, w/v)	Ethanol conc. (%, w/v)	Theoretical yield[a] %	Time (h)	Yeast species	Temp. (°C)	Fermentation condition	Reference
21.8	9.2	78.6	48	*S. cerevisiae* 2541	33	Anaerobically, static batch	[32]
25[b]	11.7	96.1	108	*S. cerevisiae* TISTR	30	Static fed-batch	[33]
6.89	3.3	93.2	11	*S. cerevisiae* CICC	37	Immobilized cells, bioreactor	[30]
19.9	9.9	93.8	36	*S. cerevisiae* Y-PEG0701	30	Static/anaerobic batch	[34]
28[b]	12.1	100.0	60	*S. cerevisiae* NP01	30	Very high gravity, static batch	[13]
11.7	5.0	83.5	24	Ordinary baker's yeast	30	Stirred, batch	[35]
20	9.0	94.1	48	*S. cerevisiae* CFTR 01	30	Static batch	[10]
24	10.5	98.3	36	*S. cerevisiae* NP01	30	Agitated, batch	[31]

[a] Theoretical yield is based o the maximum 0.51 g ethanol/g glucose.
[b] Adjusted by adding surcrose.

Table 2. Comparison of compositions of different sorghum bagasse samples

Component	[36][a]	[36][b]	[37]	[35]	[41]	[42]	[39]	[47]	[32][d,e]	[32][f]
Glucan	27.3 ± 0.4	38.3 ± 0.3	45.0	36.3	35.0 ± 0.7	40.4	34.0 + 1.0	35.1 ± 0.8	12.3	27.8
Xylan	13.1 ± 0.1	18.2 ± 0.3	NA	25.6	NA	NA	20.0 + 1.0	19.2 ± 0.6	NA	NA
Aarabinan	1.4 ± 0.0	1.8 ± 0.1	NA	2.0	NA	NA	NA	4.8 ± 0.7	NA	NA
Hemicellulose[c]	14.5 ± 0.1	20.2 ± 0.4	28.0	27.7	24.4 ± 1.1	35.5	24.0 + 1.0	24.0 ± 0.7	9.7	21.9
Lignin	14.3 ± 0.2	19.7 ± 1.1	22.0	18.6	19.0 ± 2.1	NA	20.0 + 1.0	25.4 ± 0.6	8.8	19.9

[a] Unwashed bagasse.
[b] Washed bagasse.
[c] Hemicelluose is the sum of xylan and arabinan.
[d] Moisture content of 55.8%.
[e] Composition analysis followed the procedure listed in Aravantinos.
[f] Calculated values assuming 0% moisture content.

2.1.2. Producing Ethanol from Sorghum Bagasse

After juice is expressed, the left biomass is often referred to as sorghum bagasse. Unwashed bagasse contains soluble sugars and proteins that can be extracted by water and non-soluble components that are mainly glucan, xylan, arabinan, and lignin. To analyze the composition of the sorghum bagasse in terms of contents of cellulose, hemicellulose, and lignin, some studies [35-37] adopt the NREL protocol for determining structural carbohydrates and lignin in biomass [38] while others [32, 39-41] used procedures published elsewhere. Though the varieties and sources of sweet sorghum are different, the composition of sorghum bagasse varies slightly. Generally speaking, the contents of cellulose, hemicelluose, and lignin are in the range of 28-45%, 18-36%, and 19-25%, respectively (Table 2). The low contents of the three polymers reported by Mamma [32] are due to high moisture content (55.8%) of the wet biomass. If this water content is taken into consideration, then the composition of that sorghum bagasse fits into the normal range.

The high carbohydrate content of sorghum bagasse makes it an excellent candidate as a biomass feedstock for biofuel production. Producing ethanol directly from sorghum bagasse is certainly attractive, but the cellulose conversion of untreated bagasse is rather low, 16% and 23.2% as reported by Sipos [35] and Dogaris [42], respectively. Therefore, to achieve high glucan and xylan conversion and high ethanol yield, effective treatment train needs to be identified to disrupt the recalcitrant bagasse structure, hydrolyze the pretreated material, and convert the hydrolysates to various fuels. Several research groups have already reported their work on sorghum bagasse pretreatment, such as sulfuric acid [40, 41], hydrochloric acid [39], phosphoric acid [43, 44], steam [35], dilute ammonia hydroxide [37], ammonia fiber explosion [36], hot water [42], and microbial approach [45].

Reports from one research group have particularly focused on using sorghum bagasse for generating value-added chemicals, such as xylose for xylitol [39, 40, 46, 47] and furfural [43]. But the pretreatment process can definitely be borrowed for ethanol production once the sugars are released. One pretreatment scheme of using 6% hydrochloric acid seems very promising [39]. After pretreatment at this acid concentration for 83 min at atmospheric temperature, 95.5% of xylose was freed. Under this treatment condition, a low percentage of glucose was also liberated together with a relatively low concentration of furfural but a high concentration of acetic acid. It may be worth the effort to evaluate lignin removal during the acid pretreatment step and determine glucose yield after hydrolyzing the pretreated material.

Other groups have been concentrating on producing ethanol from sorghum bagasse. As shown in Table 3, different techniques give different glucan and xylan conversions as well as different ethanol yields. In a study conducted by Dogaris, when sterilized sorghum bagasse was supplemented to the cellulolytic and hemicellulolytic enzyme system produced by *Neurospora crassa*- a mesophilic fungus, glucan and xylan conversions were reported as 14.7% and 20.3%, respectively [45]. For the same sorghum bagasse, when the pretreatment condition was 210°C for 20 min and the synergistic enzyme system produced by *Fusarium oxysporum* and *Neurospora crassa* was used, a higher glucan conversion of 38% was achieved [42]. However, from these two studies, one cannot make the conclusion on whether harsher pretreatment conditions or the different enzyme system contributes most to the increased cellulose degradation.

Other studies have adopted commercial enzymes for hydrolyzing pretreated bagasse and have reported a much higher glucan and xylan conversion. Among different physical and chemical pretreatment strategies, steam pretreatment at 200°C for 5 min plus impregnation with 2% SO_2 resulted in the highest glucan conversion of 92% [35]. Pretreatment at 190°C for 10 min achieved a similar result with 89% of cellulose being hydrolyzed. Treatments at lower temperatures and with shorter residence times led to lower glucan conversion efficiencies. Xylan conversion was not reported by this study. But the ethanol yield was estimated to be 80-90% though detailed calculation was not provided.

The highest ethanol yield in terms of g ethanol/100 g bagasse is reported by Salvi [37]. In this study, dilute ammonia hydroxide (28% v/v) was used to pretreat unwashed but dried sorghum fiber. Pretreatment was conducted under one condition: sorghum fiber, ammonia, and water at a ratio of 1:0.14:8 at 160°C for 1 h with 140-160 psi pressure. After pretreatment, the untreated and pretreated biomass was subject to hydrolysis at 10% solid loading with two different enzyme doses, either full or half strength. At full strength, higher cellulose (84%) and hemicellulose (73%) conversions were observed, which were significantly higher than those of untreated samples, 38% and 14.5% for cellulose and hemicellulose, respectively. Following hydrolysis, *S. cerevisiae* D_5A was used to ferment the sugars to ethanol. A maximum of 25 g ethanol/100 g dry bagasse was achieved with hydrolysates derived from full strength enzyme concentrations. For untreated samples, the ethanol yield was 10 g/100 g dry bagasse.

The ethanol yield in terms of theoretical value for this study is 84%, which is not the highest among the list (Table 3). This yield is calculated by dividing 25 g/l (final ethanol concentration) by ethanol concentration theoretically available when assuming: 1) all cellulose in the biomass is converted to glucose and 2) xylose cannot be used by yeast *S. cerevisiae* D_5A. This kind of computation is different from others. For example, in the study conducted by Li [36], the percentage of ethanol yield is calculated only for the fermentation step through dividing actual ethanol concentration by theoretically available ethanol content from glucose and xylose which are utilized by the yeast. Using the second approach, the ethanol yield from Salvi's study is calculated as 104% (25 g/l divided by 0.51 × 47 g/l (glucose concentration in the broth)) which is the highest among all investigations on sorghum bagasse for ethanol. This high ethanol yield probably benefits from the high cellulose content in the bagasse- 45%, the highest among all bagasse samples tested (Table 3).

Table 3. Comparison of studies investigating ethanol production from sorghum bagssse

Pretreatment method	Optimal condition	Enzymes for hydrolysis	Glucan conversion (%)	Xylan conversion (%)	Ethanol yield (g/100 g dry bagasse)	Fermentation condition	Ethanol yield (%)	Fermenting microbial species	Reference
AFEX[a]	120% moisture content, 30 min, 2:1 NH$_3$/biomass loading, 140°C	Spezyme CP, Multifect xylanase, Novozyme 188	67	76	11	Shaking flask, 30°C	82.2	S. cerevisiae 424A (LNH-ST)	[36]
AFEX[b]			68	88	15.9		96.9		[37]
Dilute ammonia	Sorghum/NH$_3$/water: 1:0.14:8 160°C, 1 h, 140-160 psi	Spezyme CP and Novozyme 188	84	73	25	Shaking flask, 30°C	84[d]	S. cerevisiae D5A	[41]
Dilute sulfuric acid	20% moisture content, 1% H$_2$SO$_4$, 121°C, 1 h	Celluclast and Novozyme 188	72	84.8	13.3	SSF, fed-batch, 32°C	NA	S. cerevisiae Ethanol Red	[35]
Steam pretreatment	200°C, 5 min	Celluclast and Novozyme 188	92	NA	NA	Stirred bottle, 30°C	80-90	Ordinary baker's yeast	[44]
Dilute phosphoric acid	120°C, H$_3$PO$_4$, 80 g/l, 80 min, solid loading: 10%	Cellualase	NA	NA	14.7	Shaking flask, SSF, 37°C	NA	S. cerevisiae	[42]
Hydrothermal treatment	210°C, 20 min	Produced by F. oxysporum and N. crassa[c]	38	NA	NA	NA	NA	NA	[45]
Sterilization	NA	Produced by N. crassa	14.7	20.3	4.7	Shaking flask, 30°C	NA	N. crassa	

[a] Without wash stream.
[b] With wash stream.
[c] F: *Fusarium*; N: *Neurospora*.
[d] Yield from the publication. A different yield as 104% is provided in the text.

The second highest ethanol yield of 15.9 g ethanol per 100 g bagasse comes from the study where sorghum bagasse was pretreated by ammonia fiber expansion (AFEX) [36]. The optimal condition for sorghum bagasse pretreatment using AFEX was identified as 120% moisture content, 30 min, 2:1 ammonia to biomass ratio, and 140°C. Under this optimal condition, maximal glucan and xylan conversions of 68% and 88% were obtained when the wash stream was added to hydrolyzed bagasse during the enzymatic hydrolysis step. The wash stream was the solution from bagasse wash with water. Using genetically engineered yeast *S.cerevisiae* 424A (LNH-ST) which can ferment xylose, but slowly as shown in this study, the ethanol yield of 96.9% was attained.

Some studies in literature which have dealt with different feedstocks and pretreatment methods, such as corn stover by ammonia recycle [48], wheat straw by wet oxidation [49], and waste house wood by dilute acid [50], have indicated the need for supplying nutrients to accomplish successful fermentation. But, Salvi and Li's investigations have demonstrated that hydrolyzed sorghum bagasse pretreated by either dilute ammonia or AFEX can be directly fed to yeast for ethanol production without conditioning, washing, or adding extra nutrients. Research by Mehmood et al., which used dilute sulfuric acid (1%) to pretreat sorghum bagasse also observed the same phenomenon [41]. These researchers stated that simultaneous saccharification and fermentation (SSF) of pretreated biomass in batch and fed-batch modes was better than separate hydrolysis and fermentation (SHF) with regard to ethanol yield. In addition, it was reported by Sipos et al., that fiber separation followed by enzymatic hydrolysis delivered a readily fermentable hydrolysates, from which 80-90% of glucose was converted to ethanol. The researchers also emphasized that separating fibers from inhibitory components was not necessary when the steam pretreatment conditions were not too harsh. Even with the harshest condition that they have tested, fiber separating only resulted in 7% more conversion efficiency [35]. Therefore, based on these four publications which appear to be the most comprehensive investigations on sorghum bagasse, using sorghum bagasse for ethanol production owns another level of advantage- a direct and simple process. Simply put, pretreated bagasse can be directly hydrolyzed after pH adjustment and hydrolysates can be directly used for yeast fermentation without needing extra supplements. However, whether this advantage is only limited to these four or to all pretreatment strategies is an unanswered question. To further maximize ethanol yield from sorghum bagasse, better mixture of enzymes and better fermenting microbes are still needed to improve the conversions of glucan and xylan and enhance xylose use and ethanol yield, respectively. This stays true to ethanol generation from biomass feedstocks other than sorghum bagasse.

2.1.3. Producing Ethanol from Whole Stalk

Unsqueezed sorghum stalk has also been tested for ethanol production. In this case, freshly harvested stalk is directly chopped to small particles and used for solid-state fermentation. According to a study disclosed by Yu [51], a maximum ethanol yield of 7.9 g ethanol/100 fresh stalk or 0.46 g ethanol/ g total sugar (91% of theoretical yield) could be achieved when the yeast cell numbers/g stalk, particle size, and moisture content were 4×10^6, 0.5-1.5 mm, and 75%, respectively. Considering the initial stalk moisture content of 71%, the ethanol yield on dry weight basis was 27.2 g ethanol/100 g dry biomass. This yield was obtained with the aid of sulfurous acid which helped maintain an anaerobic condition during yeast fermentation at 37°C. The best sulfurous acid concentration was determined to be 30 mg/100 g fresh stalk. Higher concentration of this compound resulted in inhibitory effects on yeast growth.

In another publication by Wang and Liu, fermenting both of the juice and fresh stalk was studied at 30°C by using a mutated yeast strain *S. cerevisiae* Y-PEG0701 [34]. The juice was fermented through conventional submerged ethanol fermentation while the ground fresh stalk powder was fermented through solid-state fermentation. In terms of the juice, the maximum ethanol concentration was 9.94% (w/v) which was equal to 9.35 g ethanol/100 g juice leading to 93.8% of the theoretical yield. With regard to whole stalk, a 9.04% (w/w) ethanol concentration was observed, which was the same as 25.8 g ethanol/100 g dry stalk while the ethanol yield reached 94.5% of theoretical value.

Based on the mass balance presented by these two authors, from one ton of fresh sorghum stalk, 69.7 kg and 76.5 kg of ethanol can be produced from the juice and dried stem powder, respectively. Thus, the difference between these two numbers, 6.8 kg ethanol per ton of sorghum stem is due to conversion of glucan and xylan during solid-state fermentation. Assuming that the fresh stalk contains 69.1% of water- an average moisture content of 67.2% by Mamma [32] and 71% by Yu [51], the ethanol yield from complex carbohydrates on a dry weight basis will be 2.2 g ethanol/100 g dry stalk, which is 8.5% of the total ethanol yield from whole stalk. Thus, during solid-state fermentation, ethanol is mainly produced from soluble sugars in the dried sorghum stalk. From the perspective of maximizing ethanol production from sweet sorghum, juice and bagasse definitely need to be fermented separately.

Report by Mamma et al., in 1995 [32] provides such a supportive evidence. In this study, sweet sorghum was pressed to form juice and press cake. The juice was fermented and resulted in an ethanol yield of 4.8 g ethanol/ 100 g fresh stalk. The press cake was autoclaved and then converted to ethanol by a mix culture of *Fusarium oxysporum* and *S. cerevisiae*. Combining the yields from these two components, the total ethanol yield was 9.9 g ethanol/100 g fresh stalk, which is higher than the 7.65 g ethanol/100 g fresh stalk given by Wang and Liu [34]. Again, if considering the 67.2% moisture content, on a dry weight basis, the ethanol yield is 30.2 g ethanol/100 g dry stalk which is the highest among the three studies on fresh stalk use. However, if using today's commercial cellulases, much higher ethanol yield should be achieved.

2.2. Lipid Production from Sorghum

In contrast to intensive studies on ethanol production from sweet sorghum, utilization of sorghum for lipid production has not been investigated much. In terms of sorghum bagasse for this purpose, no studies can be found in literature. With regard to juice use for the same application, only three studies can be located. One publication reported that sweet sorghum juice could be used for lipid production by heterotrophic microalga *Chlorella protothecoides* [11]. Major fatty acids accumulated in the cells were oleic acid (18:1), and linoleic acid (18:2) which are suitable for biodiesel production. Another short communication demonstrated that crushed sorghum stems supported growth and lipid production of one fungal species, *Mortierella isabellina* [52]. The detailed growth characteristics, however, were not provided except an oil efficiency of 11 g/100 g dry sorghum and the major fatty acids as palmitic (16:0) and oleic acid (18:1).

Table 4. Growth comparison between *S. limacinum* SR21 and *C. protothecoide* on sorghum juice

Parameter	Unit	*S. limacinum* SR21 [29]	*C. protothecoides* [11]
Maximum dry weight	g/l	9.38 ± 0.25	5.10
Biomass productivity	g/l/day	1.87 ± 0.05	1.02
Specific growth rate	/day	0.91 ± 0.02	NA
Growth yield	g/g	0.45 ± 0.21	NA
Substrate uptake rate	g/l/day	4.76 ± 0.1	NA
Lipid content	%	73.4 ± 2.3	52.50
Lipid productivity	g/l/day	1.38 ± 0.02	0.54
DHA productivity	g/l/day	0.47 ± 0.02	NA

The third study which is also the most recent demonstrates that sorghum juice can be an excellent substrate for high value lipid generation. When sorghum juice was fed to one marine microalga, *Schizochytrium limacinum* SR21, significant amount of docosahexeanoic acid (DHA, 22:6n-3) was accumulated [29]. DHA, a major component in fish oil, is an important omega-3 polyunsaturated fatty acid (n-3 PUFA) that has been shown to have beneficial effects on preventing human cardiovascular diseases, cancer, schizophrenia, and Alzheimer's disease. Growth of *S. limacinum* SR21 on sorghum juice has been found to be dose-related. Among the juice dose of 100%, 75%, 50%, and 25%, 50% of juice content gave the highest biomass density (9.38 g/l) which was similar to that from pure glucose at 48 g/l (10.9 g/l) while high juice concentrations at 75 or 100% inhibited cell growth. By monitoring concentrations of glucose, fructose, and sucrose by HPLC, it has been discovered that *S. limacinum* SR21 only utilized glucose for growth, but not fructose and sucrose. Comparing with data for *C. protothecoides* which was inoculated to juice supplemented with an invertase to hydrolyze sucrose to glucose and fructose, *S. limacinum* SR21 has much better biomass and lipid productivities (Table 4).

Lipids accumulated in *S. limacinum* SR21 can certainly be used for biodiesel production. Published by Wen's group, biodiesel prepared by direct transesterification of dry algal biomass satisfies almost all ASTM standards, such as free glycerol, total glycerol, acid number, soap content, corrosiveness to copper, flash point, viscosity, and particulate matter except water and sediment content as well as sulfur content [53]. The relatively high sulfur content could be from sulfuric acid used for the esterification reaction. Thus, *S. limacinum* SR21 is a promising microalgal strain for further study in terms of developing biodiesel and other green fuels from sorghum juice.

2.3. Hydrogen Production from Sorghum

Sorghum juice, bagasse, or the whole plant has been tested for hydrogen production through microbial processes [19, 20, 54, 55]. As described in Table 5, for the same sorghum variety, Keller, studied by one group of researchers in Greece, different microorganisms results in different hydrogen yield. Only by the indigenous microbial communities in the sorghum extract, the hydrogen yield was 0.93 mol H_2/mol glucose or 10.51 liter/ kg sorghum [19, 20]. At a similar temperature of 37°C, but with the addition of the mesophilic fibrolytic bacterium, *Ruminococcus albus*, the H_2 yield regarding these two terms were 2.61mol H_2/mol glucose and 28 liter/kg wet sorghum, respectively [56]. Therefore, *R. albus* has greater potential in fermenting hexoses and pentoses in the sorghum juice than the original microbes in the juice. Besides converting fermentable sugars to hydrogen, *R. albus* was able to degrade cellulose and hemicellulose to simple sugars in the sorghum bagasse or residue and produced a H_2 yield of 2.59 mol/mol glucose or 31 liter/kg sorghum. From the whole stalk, the hydrogen yield of 58 liter/kg sorghum was exactly the sum of those from juice and bagasse.

Hyperthermophilic *Caldicellulosiruptor saccharolyticus* is a Gram-positive, fermentative anaerobe and can grow on a wide variety of substrates including cellulose, hemicellulose, starch, pectin, pentoses, and hexoses [57-59]. When fed by air-dried and homogenized whole sorghum plant which contained leaves and stalks, *C. saccharolyticus* produced higher H_2 yield of 1.75 mol/mol glucose at 2% dry biomass content than 1.23 mol/mol glucose at 3% of dry solid concentration [21]. At 2% solid concentration, *C. saccharolyticus* released H_2 for 34 days. The fractions of sorghum plants, leaves, stems, and juice contributed to 53.2%, 39.3%, and 7.5% toward the overall H_2 production. Under the same experimental conditions, H_2 yield from sorghum was 22% less than that from wheat straw.

In another study by Panagiotopoulos [60], sorghum bagasse pretreated by NaOH was used to support growth of *C. saccharolyticus* and produce H_2. The maximal H_2 yield observed in batch experiments was 2.6 mol/mol C6 sugar while the maximal volumetric hydrogen production rate was 10.6 mmol/l-h. These values were obtained when hydrolysates of sorghum bagasse containing 10 g sugar/l was used. With similar sugar concentration, similar H_2 yield was observed for paper sludge hydrolysates [61]. When the sugar concentration in the hydrolysates was 20 g/l, less H_2 yield of 1.3 mol/mol C6 sugar and 10.2 mmol/l-h were attained. These results compare well with H_2 generation from Miscanthus on 14 g sugar/l [62].

Table 5. Comparison of hydrogen production from different sorghum varieties using different microbial processes

Country	Sorghum variety	Sorghum part	Microorganism	Temperature (°C)	H_2 yield mol/mol glucose	Theoretical yield (%)	H_2 yield liter/kg sorghum	Reference
Greece	Keller	sorghum stalk	*Ruminococcus albus*	37	3.15	78.75	58	
		sorghum extract	*Ruminococcus albus*	37	2.61	65.25	28	[56]
		sorghum residues	*Ruminococcus albus*	37	2.59	64.75	31	
Greece	Keller	sorghum extract	Indigenous microflora	35	0.93	23.25	10.51	[20]
Greece	Keller	sorghum extract	Indigenous microflora	35	0.9	22.5	10.4	[19]
Hungary	NA	whole plant	*C. saccharolyticus*	70	1.75	43.75	30.17[a]	[58]
Cyprus	Sucrosorgo 405	sorghum bagasse	*C. saccharolyticus*	72	2.6	65	10.6[b]	[60]

[a] dry biomass.
[b] mmol/l-h.

Comparing all the available results on H_2 production from sorghum, the highest H_2 yield comes from the study by Ntaikou [56] where stems stripped from leaves, chopped to a size of 20 cm, and milled to a particle size of 1-2 mm were used for fermentation in a batch mode at 37°C. On fermentable sugars, 69.9% and 69% of maximal theoretical yield were observed when cellobiose and glucose served as the substrate, respectively. On dry sorghum biomass, the calculated yield was 232 liter/kg dry sorghum based on the reported yield of 58 liter/kg wet biomass and considering a moisture content of 75%. This value is significantly higher than 30.17 liter H_2/kg dry biomass reported by Ivanova [21]. Thus, producing hydrogen from sorghum could be a simple one-step process where the whole plant without juice separation is used in the fermentation process by *R. albus* at 37°C.

3. LOGISTIC ISSUES RELATED TO SWEET SORGHUM USE

3.1. Harvest Time

Sorghum harvest timing has a profound effect on juice sugar concentration. As demonstrated by Sipos for Hungarian sweet sorghum variety "Bereny", the total sugar concentration in juice harvested on August 22 was 50 g/l which comprised 65% of fructose, 35% of glucose, and 0% sucrose. For samples taken on August 28, the total sugar concentration was 100 g/l with 13% of fructose, 22% of glucose, and 65% of sucrose. From that day on till September 26, total sugar concentrations were relatively stable with sucrose being the dominant sugar (57-72%). But starting from October, sucrose concentration decreased with time dramatically. By November 8, the total sugar concentration was 85 g/l with 42%, 44%, and 14% of fructose, glucose, and sucrose, respectively [35]. Similar observations have been reported by other research groups [63, 64]. Hence, it is critical to determine the best window for sorghum harvesting to obtain the highest sugar concentrations possible.

3.2. Storage

To use sweet sorghum as an energy crop, it has to be stored for year-round operation. Though sorghum stalks after juice expression are able to be stored in the field for 4-5 months as silage without quality comprise in northern atmosphere [4], stalks harvested in warmer regions tend to deteriorate slowly in the open field. Moreover, sorghum juice has an even poorer storability and requires immediate processing [4, 35, 65]. Generally, freshly squeezed juice spoils within 5 to 12 h at ambient temperatures due to the presence of various microorganisms in the juice (10^8/ml), such as *Leuconostoc mesenteroides*, yeasts, and nonfecal coliform bacteria. Therefore, if sweet sorghum is to be used as a raw material for fuels or chemicals, this deterioration must be prevented. More importantly, if sweet sorghum is to reach its potential as a biomass crop, the storage time after harvesting must be at least 6 months. Any short processing time will increase capital costs as oversized equipment will be needed [66].

Several techniques have been developed over the years to solve the storage challenge. Some strategies are specifically dedicated to juice preservation, such as: 1) evaporating the juice to a 60% syrup [27] or adding sodium metabisulphite to desulphurise the juice [1]. Others are developed to store the whole stalks. Examples include: 1) ensiling the chopped stalks in the presence of 0.5% formic acid which is a selective inhibitor of certain bacterial populations, including lactic acid bacteria and Clostridia, but not of yeast. Within 60 days, the original sugar content was completely preserved [65]; 2) fiber digestion in enzyme-assisted ensiling where commercial enzymes, Cellulust and Viscozyme were added for fiber degradation. This process increased the time window for bioprocessing by only 30 days, after which the sugar losses were high due to encouraged yeast fermentation [65]; 3) ensiling by adding commercial preparation of lactic acid bacteria. In this case, the sugar loss in 30 days was 28.6% [65]; 4) storing the stalks outdoors in sealed containers (anaerobic conditions) having either a CO_2 or SO_2 atmosphere or surface spraying with propionic acid or aqueous ammonia. The best sugar recovery after 200 days, 34%, was from propionic acid treatment while CO_2 and aqueous ammonia had no effects compared with controls. SO_2 treatment resulted in 19% of the original sugars being preserved [66]; and 5) storing the stalks under aerobic conditions by either drying or spraying with propionic acid. Under this condition, no sugars were left after 200 days for all different treatments [66].

Most laboratory studies store juice by freezing at -20°C and keep bagasse as dried and milled particles. However, the answer to whether these two options are good for large scale biofuel production from sweet sorghum is uncertain. It depends on a lot of factors, such as production size, equipment availability, and maintenance cost, etc.

4. Energy and Economic Analysis of Ethanol Production from Sweet Sorghum

Attempt to calculate the energy ratio (energy output/energy input) with regard to ethanol production from sweet sorghum was conducted in 1992 by Worley et al. [27]. Two scenarios were investigated. Under scenario 1, the juice was evaporated to a 60° Brix syrup, stored, and used for fermentation throughout the year to produce ethanol. Under scenario 2, the fresh juice was fermented during harvest season while the left by-product silage and cellulosic materials served as feedstock for the remainder of the year. Energy ratios of 0.9, 1.1, and 0.8 were found for sorghum with scenario 1, scenario 2, and corn, respectively. If only liquid fuel input was considered and nonliquid fuels (coal, biomass, and other forms of energy not suitable for use in transportation) were used wherever possible, the ratios were 3.5, 7.9, and 4.5 for sorghum with scenario 1, scenario 2, and corn, respectively. Therefore, using sweet sorghum for producing ethanol is more attractive and profitable compared to that of corn from the perspective of energy saving. In another word, if the price of liquid fuel rises with all other variables held unchanged, then sweet sorghum will gain an economic advantage over corn. In addition, scenario 2 provides better energy output in comparison with that of scenario 1.

Following Worley's study, many investigators have conducted economic/energy analysis for ethanol production from sweet sorghum planted at different locations. Two studies have specially focused on North China [4], and Inner Mongolia region of China [34]. Others have looked at Europe [67], Italy [68], and US [69, 70]. Since regional characteristics are incorporated into these analyses, results from these studies are not reviewed here. But it is very important to ensure that we can develop biofuels from sweet sorghum in a truly sustainable way even if there are no technical barriers.

CONCLUSION

To achieve the highest biofuel yield from sweet sorghum and make the best use of this plant, juice and bagasse need to be processed separately. Sweet sorghum juice is readily to be converted to ethanol by *S. cerevisiae* with at least 93% of theoretical yield. Converting juice to lipids is at a beginning stage of research, but should represent a promising area considering the high energy content in lipids compared with that from ethanol. Several techniques have been developed for pretreating sorghum bagasse, but the glucan and xylan conversion efficiencies are far from satisfactory. Genetically engineered yeast for glucose and xylose fermentation is available, but its performance on xylose utilization still needs to be improved. Sweet sorghum can be used for hydrogen production, but there is still large room for yield improvement. Large scale biofuel production from sweet sorghum can not be implemented without solving the logistic issues first. Energy, economic, and life cycle analyses need to be conducted to secure a sustainable future in terms of turning sweet sorghum to various fuels.

REFERENCES

[1] Nguyen MH: Suitability of sweet sorghum for alcohol in the Pacific. *Energy in Agriculture* 1984, 3:345-350.

[2] Sarath G, Mitchell R, Sattler S, Funnell D, Pedersen J, Graybosch R, Vogel K: Opportunities and roadblocks in utilizing forages and small grains for liquid fuels. *Journal of Industrial Microbiology and Biotechnology* 2008, 35:343-354.

[3] Prasad S, Singh A, Jain N, Joshi HC: Ethanol production from sweet sorghum syrup for utilization as automotive fuel in India. *Energy and Fuels* 2007, 21:2415-2420.

[4] Gnansounou E, Dauriat A, Wyman CE: Refining sweet sorghum to ethanol and sugar: economic trade-offs in the context of North China. *Bioresource Technololgy* 2005, 96:985-1002.

[5] Sakellariou-Makrantonaki M, Papalexis D, Nakos N, Kalavrouziotis IK: Effect of modern irrigation methods on growth and energy production of sweet sorghum (var. Keller) on a dry year in Central Greece. *Agricultural Water Management* 2007, 90:181-189.

[6] Cundiff JS: Potential reduction of nonstructural carbohydrate losses by juice expression prior to ensiling sweet sorghum. *Biomass and Bioenergy* 1992, 3:403-410.

[7] Billa E, Koullas DP, Monties B, Koukios EG: Structure and composition of sweet sorghum stalk components. *Industrial Crops and Products* 1997, 6:297-302.

[8] Bryan WL, Usda ARS: Solid-state fermentation of sugars in sweet sorghum. *Enzyme and Microbial Technology* 1990, 12:437-442.

[9] Almodares A, Hadi MR: Production of bioethanol from sweet sorghum: A review. *African Journal of Agricultural Research* 2009, 4:772-780.

[10] Ratnavathi CV, Suresh K, Kumar BSV, Pallavi M, Komala VV, Seetharama N: Study on genotypic variation for ethanol production from sweet sorghum juice. *Biomass and Bioenergy* 2010, 34:947-952.

[11] Gao C, Zhai Y, Ding Y, Wu Q: Application of sweet sorghum for biodiesel production by heterotrophic microalga *Chlorella protothecoides*. *Applied Energy* 2010, 87:756-761.

[12] Evans JM, Cohen MJ: Regional water resource implications of bioethanol production in the Southeastern United States. *Global Change Biology* 2009, 15:2261-2273.

[13] Laopaiboon L, Nuanpeng S, Srinophakun P, Klanrit P, Laopaiboon P: Ethanol production from sweet sorghum juice using very high gravity technology: Effects of carbon and nitrogen supplementations. *Bioresource Technology* 2009, 100:4176-4182

[14] Audilakshmi S, Mall AK, Swarnalatha M, Seetharama N: Inheritance of sugar concentration in stalk (brix), sucrose content, stalk and juice yield in sorghum. *Biomass and Bioenergy*, 34:813-820.

[15] El-Razek AMA, Besheit SY: Potential of some sweet sorghum (Sorghum bicolor L. Moench) varieties for syrup and ethanol production in Egypt. *Sugar Tech* 2009, 11:239-245.

[16] Shimazaki Y, Nagasaka K, Onda T, Ozawa M: Evaluation of bio-ethanol production from agricultural residue biomass (case study of Chuo City in Yamanashi prefecture). *Nihon Enerugi Gakkaishi/Journal of the Japan Institute of Energy* 2009, 88:1002-1008.

[17] Matsika E, Yamba FD: Harnessing sweet sorghum for bioenergy in Zambia. *Renewable Energy for Development* 2006, 19:6-8.

[18] Barbanti L, Grandi S, Vecchi A, Venturi G: Sweet and fibre sorghum (*Sorghum bicolor* (L.) Moench), energy crops in the frame of environmental protection from excessive nitrogen loads. *European Journal of Agronomy* 2006, 25:30-39.

[19] Antonopoulou G, Gavala HN, Skiadas IV, Angelopoulos K, Lyberatos G: Biofuels generation from sweet sorghum: Fermentative hydrogen production and anaerobic digestion of the remaining biomass. *Bioresource Technology* 2008, 99:110-119.

[20] Antonopoulou G, Gavala HN, Skiadas IV, Lyberatos G: Influence of pH on fermentative hydrogen production from sweet sorghum extract. *International Journal of Hydrogen Energy* 2010, 35:1921-1928.

[21] Ivanova G, Rákhely G, Kovács KL: Thermophilic biohydrogen production from energy plants by Caldicellulosiruptor saccharolyticus and comparison with related studies. *International Journal of Hydrogen Energy* 2009, 34:3659-3670.

[22] Liu SY, Lin CY: Development and perspective of promising energy plants for bioethanol production in Taiwan. *Renewable Energy* 2009, 34:1902-1907.

[23] Li SZ, Chan-Halbrendt C: Ethanol production in China: Potential and technologies. *Applied Energy* 2009, 86:S162-S169.

[24] Guo Y, Hu SY, Li YR, Chen DJ, Zhu B, Smith KM: Optimization and analysis of a bioethanol agro-industrial system from sweet sorghum. *Renewable Energy* 2010, 35:2902-2909.

[25] Zhang C, Xie G, Li S, Ge L, He T: The productive potentials of sweet sorghum ethanol in China. *Applied Energy* 2010, 87:2360-2368.

[26] de Vries SC, van de Ven GWJ, van Ittersum MK, Giller KE: Resource use efficiency and environmental performance of nine major biofuel crops, processed by first-generation conversion techniques. *Biomass and Bioenergy* 2010, 34:588-601.

[27] Worley JW, Vaughan DH, Cundiff JS: Energy analysis of ethanol production from sweet sorghum. *Bioresource Technology* 1992, 40:263-273.

[28] Zhao YL, Dolat A, Steinberger Y, Wang X, Osman A, Xie GH: Biomass yield and changes in chemical composition of sweet sorghum cultivars grown for biofuel. *Field Crops Research* 2009, 111:55-64.

[29] Liang Y, Sarkany N, Cui Y, Yesuf J, Trushenski J, Blackburn JW: Use of sweet sorghum juice for lipid production by *Schizochytrium limacinum* SR21. *Bioresource Technology* 2010, 101:3623-3627.

[30] Liu R, Shen F: Impacts of main factors on bioethanol fermentation from stalk juice of sweet sorghum by immobilized *Saccharomyces cerevisiae* (CICC 1308). *Bioresource Technology* 2008, 99:847-854.

[31] Khongsay N, Laopaiboon L, Laopaiboon P: Growth and batch ethanol fermentation of *Saccharomyces cerevisiae* on sweet sorghum stem juice under normal and very high gravity conditions. *Biotechnology*, 9:9-16.

[32] Mamma D, Christakopoulos P, Koullas D, Kekos D, Macris BJ, Koukios E: An alternative approach to the bioconversion of sweet sorghum carbohydrates to ethanol. *Biomass and Bioenergy* 1995, 8:99-103.

[33] Laopaiboon L, Thanonkeo P, Jaisil P, Laopaiboon P: Ethanol production from sweet sorghum juice in batch and fed-batch fermentations by *Saccharomyces cerevisiae*. *World Journal of Microbiology and Biotechnology* 2007, 23:1497-1501.

[34] Wang F, Liu C-Z: Development of an Economic Refining Strategy of Sweet Sorghum in the Inner Mongolia Region of China. *Energy and Fuels* 2009, 23:4137-4142.

[35] Sipos B, Reczey J, Somorai Z, Kadar Z, Dienes D, Reczey K: Sweet sorghum as feedstock for ethanol production: enzymatic hydrolysis of steam-pretreated bagasse. *Applied Biochemistry and Biotechnology* 2009, 153:151-162.

[36] Li B-Z, Balan V, Yuan Y-J, Dale BE: Process optimization to convert forage and sweet sorghum bagasse to ethanol based on ammonia fiber expansion (AFEX) pretreatment. *Bioresource Technology*, 101:1285-1292.

[37] Salvi D, Aita G, Robert D, Bazan V: Ethanol production from sorghum by a dilute ammonia pretreatment. *Journal of Industrial Microbiology and Biotechnology*, 37:27-34.

[38] Sluiter A, Hames B, Ruiz R, Scarlata C, Sluiter J, Templeton D, Crocker D: Determination of structural carbohydrates and lignin in biomass. *NREL, Golden, CO* 2004.

[39] Herrera A, Téllez-Luis SJ, González-Cabriales JJ, Ramírez JA, Vázquez M: Effect of the hydrochloric acid concentration on the hydrolysis of sorghum straw at atmospheric pressure. *Journal of Food Engineering* 2004, 63:103-109.

[40] Simón JT-L, José AR, Manuel V: Modeling of the hydrolysis of sorghum straw at atmospheric pressure. *Journal of the Science of Food and Agriculture* 2002, 82:505-512.

[41] Mehmood S, Gulfraz M, Rana NF, Ahmad A, Ahring BK, Minhas N, Malik MF: Ethanol production from *Sorghum bicolor* using both separate and simultaneous saccharification and fermentation in batch and fed batch systems. *African Journal of Biotechnology* 2009, 8:2857-2865.

[42] Dogaris I, Karapati S, Mamma D, Kalogeris E, Kekos D: Hydrothermal processing and enzymatic hydrolysis of sorghum bagasse for fermentable carbohydrates production. *Bioresource Technology* 2009, 100:6543-6549.

[43] Vázquez M, Oliva M, Téllez-Luis SJ, Ramírez JA: Hydrolysis of sorghum straw using phosphoric acid: Evaluation of furfural production. *Bioresource Technology* 2007, 98:3053-3060.

[44] Ban J, Yu J, Zhang X, Tan T: Ethanol production from sweet sorghum residual. *Frontiers of Chemical Engineering in China* 2008, 2:452-455.

[45] Dogaris I, Vakontios G, Kalogeris E, Mamma D, Kekos D: Induction of cellulases and hemicellulases from *Neurospora crassa* under solid-state cultivation for bioconversion of sorghum bagasse into ethanol. *Industrial Crops and Products* 2009, 29:404-411.

[46] Herrera A, Téllez-Luis SJ, Ramírez JA, Vázquez M: Production of xylose from sorghum straw using hydrochloric acid. *Journal of Cereal Science* 2003, 37:267-274.

[47] Téllez-Luis SJ, Ramírez JA, Vázquez M: Mathematical modeling of hemicellulosic sugar production from sorghum straw. *Journal of Food Engineering* 2002, 52:285-291.

[48] Kim T, Lee Y, Sunwoo C, Kim J: Pretreatment of corn stover by low-liquid ammonia recycle percolation process. *Applied Biochemistry and Biotechnology* 2006, 133:41-57.

[49] Klinke HB, Olsson L, Thomsen AB, Ahring BK: Potential inhibitors from wet oxidation of wheat straw and their effect on ethanol production of *Saccharomyces cerevisiae*: wet oxidation and fermentation by yeast. *Biotechnology and Bioengineering* 2003, 81:738-747.

[50] Okuda N, Ninomiya K, Takao M, Katakura Y, Shioya S: Microaeration enhances productivity of bioethanol from hydrolysate of waste house wood using ethanologenic *Escherichia coli* KO11. *Journal of Bioscience and Bioengineering* 2007, 103:350-357.

[51] Yu J, XuZhang, Tan T: Ethanol production by solid state fermentation of sweet sorghum using thermotolerant yeast strain. *Fuel Processing Technology* 2008, 89:1056-1059.

[52] Economou CN, Makri A, Aggelis G, Pavlou S, Vayenas DV: Semi-solid state fermentation of sweet sorghum for the biotechnological production of single cell oil. *Bioresource Technology* 2010, 101:1385-1388.

[53] Johnson MB, Wen Z: Production of biodiesel fuel from the microalga *Schizochytrium limacinum* by direct transesterification of algal biomass. *Energy and Fuels* 2009, 23:5179-5183.

[54] Saraphirom P, Reungsang A: Optimization of biohydrogen production from sweet sorghum syrup using statistical methods. *International Journal of Hydrogen Energy* 2010, In Press, Corrected Proof.

[55] Ntaikou I, Gavala HN, Lyberatos G: Application of a modified Anaerobic Digestion Model 1 version for fermentative hydrogen production from sweet sorghum extract by *Ruminococcus albus*. *International Journal of Hydrogen Energy*, 35:3423-3432.

[56] Ntaikou I, Gavala HN, Kornaros M, Lyberatos G: Hydrogen production from sugars and sweet sorghum biomass using *Ruminococcus albus*. *International Journal of Hydrogen Energy* 2008, 33:1153-1163.

[57] Rainey FA, Donnison AM, Janssen PH, Saul D, Rodrigo A, Bergquist PL, Daniel RM, Stackebrandt E, Morgan HW: Description of *Caldicellulosiruptor saccharolyticus* gen. nov., sp. nov: an obligately anaerobic, extremely thermophilic, cellulolytic bacterium. *FEMS microbiology letters* 1994, 120:263-266.

[58] Ivanova G, Rákhely G, Kovács KL: Hydrogen production from biopolymers by *Caldicellulosiruptor saccharolyticus* and stabilization of the system by immobilization. *International Journal of Hydrogen Energy* 2008, 33:6953-6961.

[59] van de Werken HJG, Verhaart MRA, VanFossen AL, Willquist K, Lewis DL, Nichols JD, Goorissen HP, Mongodin EF, Nelson KE, van Niel EWJ: Hydrogenomics of the extremely thermophilic bacterium *Caldicellulosiruptor saccharolyticus*. *Applied and Environmental Microbiology* 2008, 74:6720.

[60] Panagiotopoulos IA, Bakker RR, de Vrije T, Koukios EG, Claassen PAM: Pretreatment of sweet sorghum bagasse for hydrogen production by *Caldicellulosiruptor saccharolyticus*. *International Journal of Hydrogen Energy*, 35:7738-7747.

[61] Kádár Z, de Vrije T, van Noorden G, Budde M, Szengyel Z, Réczey K, Claassen P: Yields from glucose, xylose, and paper sludge hydrolysate during hydrogen production by the extreme thermophile *Caldicellulosiruptor saccharolyticus*. *Applied Biochemistry and Biotechnology* 2004, 114:497-508.

[62] de Vrije T, Bakker RR, Budde MA, Lai MH, Mars AE, Claassen PA: Efficient hydrogen production from the lignocellulosic energy crop *Miscanthus* by the extreme thermophilic bacteria *Caldicellulosiruptor saccharolyticus* and *Thermotoga neapolitana*. *Biotechnology for Biofuels* 2009, 2:12-27.

[63] Amaducci S, Monti A, Venturi G: Non-structural carbohydrates and fibre components in sweet and fibre sorghum as affected by low and normal input techniques. *Industrial Crops and Products* 2004, 20:111-118.

[64] Hoffmann-Thoma G, Hinkel K, Nicolay P, Willenbrink J: Sucrose accumulation in sweet sorghum stem internodes in relation to growth. *Physiologia Plantarum* 1996, 97:277-284.

[65] Schmidt J, Sipocz J, Kaszas I, Szakacs G, Gyepes A, Tengerdy RP: Preservation of sugar content in ensiled sweet sorghum. *Bioresource Technology* 1997, 60:9-13.

[66] Jesberg BK, Montgomery RR, Anderson RA: Preservation of sweet sorghum biomass. *Biotechnology and Bioengineering symposium* 1983, 13:113-120.

[67] Venturi P, Venturi G: Analysis of energy comparison for crops in European agricultural systems. *Biomass and Bioenergy* 2003, 25:235-255.

[68] Monti A, Venturi G: Comparison of the energy performance of fibre sorghum, sweet sorghum and wheat monocultures in northern Italy. *European Journal of Agronomy* 2003, 19:35-43.

[69] Hallam A, Anderson IC, Buxton DR: Comparative economic analysis of perennial, annual, and intercrops for biomass production. *Biomass and Bioenergy* 2001, 21:407-424.

[70] Turhollow A: The economics of energy crop production. *Biomass and Bioenergy* 1994, 6:229-241.

Chapter 4

CHEMICAL COMPOSITION OF FORAGE SORGHUM AND FACTORS RESPONSIBLE FOR INCREASING ANIMAL PRODUCTION

Sandeep Kumar, Kaushalya Gupta and U. N. Joshi
Chemical Laboratory, College of Agriculture, CCS Haryana Agriculture University, Hisar (Haryana) India

With the increasing cattle population, the gap between demand and supply of fodder is also increasing which can be reduced through improvement of forage crops especially forage sorghum. Forage sorghum (*Sorghum bicolor* L. Moench) is grown mainly for fodder purpose during summer and *kharif* seasons in the northern states of India. Kamalak *et al.*, (2005) reported that protein up to a minimum level (7%) is essential to maintain rumen micro-flora for proper digestion as well as intake of the fodder. The productivity of the livestock depends upon good quality fodders. Animal performance is mainly dependent on three factors namely intake, digestibility and utilization. Keeping these points in mind the present chapter has been discussed under the following heads

1. Genetic variability in quality attributes
2. Effect of management practices on quality attributes
3. Effect of pathogens on quality of sorghum

Man is directly dependent upon plants for his survival because plants directly are the prime source of food, fibre and drugs. With the increasing cattle population, the gap between demand and supply of fodder is also increasing which can be reduced through improvement of forage crops especially forage sorghum. Being a dual purpose crop, the importance of Sorghum (*Sorghum bicolor* L. Moench.), hardly needs emphasis. It is a major grain crop of Central and South India, but it is extensively grown for fodder in North India during summer and kharif season.

Sorghum, (*Sorghum bicolor* L.) Moench is the world's fifth most important cereal crop, behind rice, corn, wheat and barley and the third leading crop in the U.S.A. It belongs to family Graminea, commonly called as sorghum and also known as 'durra', 'chari' or 'jowar', is a grass species, often grown in area of relatively low rainfall, high temperature and saline soil (Boursier and Lauchli, 1990). In India, it is cultivated in about 16 million hectares, across Maharastra, Madhya Pradesh, Uttar Pradesh, Haryana, Punjab, Andhra Pradesh, Tamil Nadu, Karnataka and Rajasthan. It is used for food, fodder and the production of alcoholic beverages, but extensively grown for fodder in north India during summer and *kharif* season due to its greater adaptability, high fodder yield, better palatability, quality, digestibility and various forms of utilization like green chop, Stover, silage and hay. It gives higher fodder yield even under adverse and poor management conditions. The proper care is to be taken in nutrient managements for growing sorghum crop, because it depletes soil fertility very fast. Kamalak *et al.*, (2005) reported that protein up to a minimum level (7%) is essential to maintain rumen micro flora for proper digestion as well as intake of the fodder.

The sorghum is a nutrient exhaustive crop and its productivity is low because of insufficient supply or sub-optimal use of nutrients in general and nitrogen, phosphorus, potassium and zinc in particular. Thus, a suitable cultivar and proper nutrition are very important to get higher fodder yield with best quality. Keeping these points in mind the present chapter has been discussed under following heads.

1. GENETIC VARIABILITY IN QUALITY ATTRIBUTES

1.1. Hydrocyanic Acid

A number of common plants may accumulate large quantities of prussic acid (cyanogenic compounds). Sorghums and related species readily accumulate these compounds. Hydrocyanic acid content in sorghum is heritable and subjected to modification through selection and breeding, as well as by climate, stage of maturity, stunting of plant, type of soil and fertilizer (Khatri *et al.*, 1997), and increased during stress. HCN content in at least 1,000 genotypes varied from 14 to 374 ppm, is higher in genotypes grown in summer.

At young stage (35 DAS) the concentration of HCN was maximum and after that it decreased. This drop in HCN content as the plant ages may be contributed to the fact that the prussic acid content in green plants decreases as maturity approaches. Most of the genotypes had HCN content less than the lethal dose of 200 ppm. Six genotypes, viz, S 532, SRV 11-1, IS 651-2 S, G 73-5, SSG GSS BH and EJ 167 (IS 3228 and 74295) were superior, which contained least amount of HCN i.e $\leq$ 10 ppm HCN

1.2. Protein Content and IVDMD (*In Vitro* Dry Matter Digestibility)

Protein content in single cut (SC) and multicut (MC) genotypes ranged from 5.24 to 10.06 and 4.81 to 12.47 per cent, respectively. Similarly, IVDMD ranged from 50.4 to 62.0 and 48.3 to 62.2, respectively. Protein and DDM yields (q/ha) in SC, MC dual purpose genotypes varied from 0.16 to10.63, 8.19 to 29.49 and 1.04 to 6.34., 1.00 to 100.34, 30.85 to 156.47 and 5.90 to 39.26, respectively. Considering the protein and DDM yield, the variety (UTMC 536 and SSG 59-3) was excellent as it ranked first , while among hybrids, CSH 20

MF was superior when protein yield was considered and HH 386 was superior when DDM yield was considered. Best genotypes were found to be HC 308, S 513, HH 4, FSH 9207, HH 85,

2. EFFECT OF MANAGEMENT PRACTICES ON QUALITY ATTRIBUTES

Plants in turn are affected by a number of natural phenomenon. The major environmental factor that currently reduces plant productivity is salinity (Majeed *et al.*, 2010). Salinity, due to over-accumulation of NaCl, is usually of great concern and the most injurious factor in arid and semi arid regions. It is one of the major environmental factors that limit productivity and quality of many economically important crops throughout the world (Sarifi *et al.*, 2007). Soil salinity is one of the main problems for agriculture, especially in countries where irrigation is an essential aid to agriculture (Ahloowalia *et al.*, 2004). This is mainly due to low precipitation and high transpiration causing disturbance in salt balance in the soil; this also renders ground water brackish and affects plant growth adversely (Rhoades & Loveday, 1990; Evans, 1998) and the area affected by salinity is increasing steadily (Ghassemi *et al.*, 1995). There are vast areas in Iran with salinity-affected soils and moderate saline soils occupy approximately 25.5 million ha and strong saline soils cover about 8.5 million ha. Soil salinity reduces yield production of most crops (Munns *et al.*, 2002). Soil salinity was shown to increase P, Mn and Zn and decrease K and Fe (Turan *et al.*, 2007) concentrations of plants.

Screening of large number of genotypes of a crop is necessary to identify the salt tolerant germplasm for breeding programs to evolve the salt tolerant and high yielding crop varieties.

Combination of fertilizers are required to be used to maintain the microbial ecology as well as soil health and hence production potential and quality of the crops. Application of bio-fertilizer to the soil / seed accelerates the extent of nutrient availability supplements the demand of chemical fertilizers and enhances the growth and biomass of plants. Use of bio-fertilizer is an effective component in eco-friendly production systems partially substituted for FYM and inorganic fertilizers (Kumar, *et al.*, 2005). The development of IPNS models for various forage species should be given priority to improve the growth of forage crops to improve the productivity of soil. There fore it was thought to know the sustainability and quality of forage sorghum through the integrated nutrient management of chemical fertilizers supplemented with farmyard manure and biofertilizer. Combination of fertilizers are used to maintain the microbial ecology as well as soil health and hence production potential and quality of the crops.

2.1. Effect of Management Practices on Yield and Quality Attributes

Maximum plant height (338.7 cm) and number of tillers/plant (19.1) was obtained with 120 kg N/ha. Significantly higher green as well as dry matter yield (391.25 and 121.8 q/ha) was recorded with 120 kg N/ha over other levels, treatments, on pooled mean basis. Higher green and dry matter yields were observed with nitrogen fertilization in comparison to that of phosphorus.

Gupta *et al* (2007 and 2008) reported an increase from 44.2 to 99.8 per cent for green fodder and 19.4 to 41.2 per cent for dry matter yield with the increasing application of nitrogen (from 40, 80 and 120 kg N/ha) over control, respectively. Since soils of the experimental fields were deficient in available nitrogen and medium in phosphorus, which are essential elements for growth, regeneration and development of plants, therefore, such increases were expected. Significantly increased plant heights and number of tillers/running meter row length were also observed under increasing level of nitrogen and phosphorus. Similar results have also been reported by Sheoran and Rana (2006) and Chaurasia *et al.* (2006).

2.2. HCN, Crude Protein and IVDMD

The HCN content increased with nitrogen application, where as, it decreased with subsequent increase in phosphorus and irrigation levels. Low HCN content appeared to be dominant over high HCN content. Regarding toxic constituents, HCN is higher in summer than in kharif season. The HCN content ranged from 25 to 710 ppm on dry weight basis and it was found to be maximum at 30 days stage.

Quality traits were also influenced with fertility levels. Protein and DDM yields due to nitrogen fertilization were superior over control treatment. Protein and DDM yields ranged from 3.36 to 5.89 and 40.86 to 60.42 q/ha, respectively. Similar trend was observed in the case of phosphorus fertilization. Norton (2003) and Kamalak *et al.*, (2005) reported that the utilization of forage depends on the micro flora build up in the rumen which in turn would require a threshold value (7%) of protein in the forage.

Protein synthesis is an important event which occurs early during germination. The highest protein yield and DDM was recorded in variety SSG-59-3 followed by CSH 20 MF. Protein content and yield decreased in second cut as compared to that of first cut. There was significant increase in protein and DDM yield with increase in fertilizer application, ranged from 7.85 to 14.06 and 38.09 to 66.10 q/ha, respectively. The variety CSV-21 F ranked first in protein and DDM yield. Protein and DDM yield ranged from 7.78 to 12.80 and 34.65 to 61.19 q/ha, respectively.

The highest protein yield and DDM was recorded at 75% RDF+ 25% N through FYM. Protein content and yield decreased in second cut as compared to that of first cut in all the treatments. Protein and DDM yields (q/ha) ranged from 3.36 t0 5.89 and 40.86 to 60.42 respectively under different N and P (Gupta *et al*, 2008).

Gupta *et al.*,(2007) reported the range of protein and IVDMD percentage among different treatments from 4.60 to 6.35 and 35.0 to 43.8, respectively. The protein and DDM yield ranged from 5.81 to 9.31 and 32.94 to 67.21 q/ha, respectively. The highest protein yield and DDM was recorded at 100% RDF+ 25 Kg $ZnSO_4$/h and 100% RDF+ 15 Kg $ZnSO_4$/h treatments respectively.

2. 3. Structural Carbohydrate Content

Lignin play a role as performed resistance factor; induced lignifications have been proposed as an active resistance mechanism of plants against stress, at the site of wounding, lignin formation has been observed apparently to strengthen the cell wall at the location of damage.

The ADF, NDF, Cellulose, Hemi-cellulose and lignin percentage decreased in second cut as compared to that of first cut in all the treatments. Their average percentage varied from 42.10 to 44.65, 73.60 to 76.66, 7.13 to 7.7, 31.15 to 33.45 and 4.73 to 5.55 respectively. Maximum amount of ADF, NDF and Cellulose were recorded at 100% RDF through inorganic fertilizer, 100% N through FYM and 50% RDF+25% N through FYM+*Azospirillum* respectively.

2.4. Non- Structural Carbohydrates

Carbohydrates in the cell fulfill dual functions. On the one hand it acts in the capacity of an osmotic agent; while on the other hand, it provides energy (ATP), reducing power (NADPH) and carbon skeleton for biosynthesis.

These compatible solutes include mainly proline, glycine, betaine and sugars (Ghoulam *et al.*, 2002 and Khan *et al.*, 2000) which accumulate in plants under salt stress and their major functions had been reported to be osmotic adjustment, carbon storage and radical scavenging.

As explained by Giorgini and Suda (1990), the higher level of soluble sugars detected is probably necessary for the turgor and growth of embryonic axes during emergence. In addition, Singh (2004) proved that a greater accumulation of sugar lowers the osmotic potential of cells and reduces loss of turgidity in tolerant genotypes. The other possible role of sugar may be as a readily available energy source.

Accordingly, it may be concluded that high soluble sugars play an important role in turgor maintenance and proline is regarded as a source of energy, carbon and nitrogen for recovering tissues, so it increased under water stress levels (Blum and Ebercon, 1981).

2.5. Mineral Content

In view of the importance of the mineral content of forages in animal nutrition, comprehensive analysis of most tropical forages reveals that these are lower in mineral contents than the temperate speicies. Maximum amount of Zn, Fe and K was recorded at 50% RDF+*Azospirillum* and 50% RDF+25% N through FYM+ *Azospirillum* respectively. Amount of Cu, Zn , Mn and K decreased in second cut as compared to that of first cut in all the treatments with few exception. Total Cu, Zn, Fe, Mn , Na and K contents of two cuts varied from 57.75 to 93.25 ppm, 40.70 to 94.30 ppm, 672.40 to 849.40 ppm, 70.25 to 109.40 ppm, 0.060 to 0.081 % and 3.78 to 4.26%, respectively.

Total Cu, Zn, Fe, Mn , Na and K contents of two cuts varied from 35.20 to 48.95 ppm, 44.95 to 63.90 ppm, 439.0 to 524.60 ppm, 68.80 to 82.65 ppm, 0.108 to 0. 120% and 4.32 to 4.46 % , respectively. These variations may be due to agroclimatic condtions. Maximum amount of Cu, Zn, Mn and K was recorded at 50% RDF+25% N through FYM+ *Azospirillum*. Amount of Cu, and Mn increased in second cut as compared to that of first cut in all the treatments.

Minerals uptake by sorghum plant was also influenced under different level of N and P application. Mn, Fe, Zn and Cu contents (ppm) ranged from 25.59 to 28.68, 227.31 to 298.41, 37.50 to 43.38 and 11.75 to 13.38 ppm, respectively. Highest amount of Mn (28.68 ppm) and Fe (298.41ppm) was observed with 120 kg and 40 kg N/ha treatments, respectively. Highest amount of Cu (13.38ppm) and Zn (43.38 ppm) was observed in 20 kg P_2O_5/ha treatments .Variation in Na and K content were also observed (Gupta *et al.*, 2008). Higher amount of phosphorus in roots and stubbles of MC genotypes has been observed which should play a definite role in regeneration.

Highest amount of Fe (298.4ppm) and Mn (28.68 ppm) was observed in 40 kg and 120 kg N/ha treatments, respectively. Highest amount of Cu (13.38ppm) and Zn (43.38 ppm) was observed in 20 kg P_2O_5/ha treatments. Variation in Na and K content were also observed (Gupta *et al.*, 2008). (UPFS-37, UPFS-38, SRF-239, SU-658 and HC 308) responded significantly up to 80 kg N/ ha (Singh and Sumeriya, 2005). The phosphorus application is important due to its key role in root development, regeneration, energy transformation and metabolic processes of plants. The role of nutrients for increasing fodder production of good quality is vital. Therefore, required efforts are needed to increase the productivity as well as quality of forage crops by adopting different nutrient management studies.

2.6. Phenolic Compounds

Phenolic compounds plays important role among plants. These are grouped under specific low molecular weight antimicrobial compound called phytoanticipins which are present as pre-infectional and pre-exixting factors and are activated and accumulated to toxic levels after any stress. The phenolics and oxidative enzymes particularly the peroxidases are involved in defense mechanism against stem borer.

Singh (2004), found that tolerant genotypes of chickpea (*Cicer arietinum*) showed a higher level of total phenols, whereas, a significant reduction was observed in susceptible genotypes which is the same as the results of Dostanova *et al.* (1979); Latha *et al.* (1989). The same author confirmed also that phenol accumulation in tolerant genotypes could be a cellular adaptative mechanism for scavenging the free radicals of oxygen and preventing sub cellular damage during stress.

2.7. Tannin Content

Tannin content affects the palatability and digestibility in forage sorghum. It varied from 0.12 - 6.2% as catechin equivalent on dry weight basis in sorghum grains, while in fodder plant tannin content varied from 0.3- 2.9%. Tannin content in grains of sweet varieties was low while in non sweet varieties , it was high. Forage sorghum varieties with higher tannin content are lower in digestibility than low tannin genotypes. Lines like PJ7R, *S.roxburghii*, IS 3247, HC 136, SPV 98, S 120, S 109, S162, HC 260 and HC 171 have been confirmed to have low tannin content. These lines are being used extensively in forage sorghum breeding programmes.

Tannins are polymeric flavanols formed by condensation of monomeric units such as flavan-3-ols and flavan-3,4-diols (Foo *et al.*, 1996). In *Lotus* species used as forages, tannin content in the range of 1 to 5 mg/g DM is beneficial, in that it prevents cattle bloating and intestinal invasion by parasites (Niezen *et al.*, 1995; Waghorn and Shelton, 1997; Otero and Hidalgo, 2004).

In addition, Tannins have been proposed to play a role in the interactions between plants and microorganisms, either pathogenic or mutualistic, as well as in plant responses to abiotic stresses (Panckurst and Jones, 1979; Barry and Manley, 1986; Dixon and Paiva, 1995; Gebrehiwot *et al.*, 2002; Reinoso *et al.*, 2004; Paolocci *et al.*, 2005).

Tannins have both beneficial and adverse effects. Beneficial effects of tannins include suppression of bloat (Jones *et al.* 1973) and protection of dietary proteins in the rumen. The adverse effects of tannins are associated with their ability to bind with dietary proteins, carbohydrates and minerals.

For meeting growing need of good quality live stock and poultry feed, sorghum [*Sorghum bicolor* (L) Moench] is one of the important forage crop mainly grown for its grain and fodder purposes due to its excellent growing habits, high yielding potential, better palatability, digestibility and better nutritive value in summer and monsoon season in northern India. Combination of fertilizers are used to maintain the microbial ecology as well as soil health and hence production potential and quality of the crops.

3. EFFECT OF PATHOGENS ON QUALITY OF SORGHUM

The biotic factors (insect, disease, and weeds), management, harvesting method and storage techniques combine to cause crop and nutrient losses as high as 25% of production. Foliar diseases caused by *Helminthosporium, Cercospora, Ramulispora* and *Gleocercospora* reduce the forage quality by inducing higher levels of undesirable constituents and by reducing the amount of desirable ones (Arora, *et al.*, 1985). Also IVDMD is reduced by 5.1% due to mite infestation and affected mainly by fibre component i.e. NDF and ADF (Arora *et al.*, 1987). Similarly, the decrease in IVDMD by about 13- 16% at more than 50% infestation is primarily due to an increase in fibre components and lignin contents (Singh *et al.*, 1995).

The phosphorus, iron and manganese content increased with diseased intensity, while Zn content significantly decreased (Arora *et al.*, 1985). Zn content is also negatively correlated with tannin content (-0.844 **).

Arora *et al.*, (1985) and Luthra *et al.*, (1997) had repoted tha supplementation of Zn and Na in the diet of animals being fed with sorghum varieties is essential for better growth, development and reproduction. It is suggested that selecting plants for higher digestibility will not lead to improved animal performance if the plan is seriously deficient in an essential element or has toxic constituents or both as low levels of essential elements frequently limit animal productivity.

REFERENCES

Ahloowalia, B.S., Meluzynski, M., Nichterlein, K. (2004). Global impact of mutation-derived varieties. *Euphytica* 135: 187-204.

Arora, S. K., Luthra, Y.P. and Joshi, U.N. 1985. Mineral status of forage sorghum leaves as affected by foliar diseases . In Proc. Of the Symp. On " *Recent Advances in Mineral Nutrition*" held at CCS HAU ,Hisar, 29th to 31st Oct. , Mandokhot , V.M. (ed.), pp 227-231

Arora, S. K., Luthra, Y.P. and Joshi, U.N. 1987. Reduction in the quality of sorghum fodder due to mite,*Oligonychus indicus* (Hirst) infestation. *Ann. Bot.* 3 :54-56.

Barry, T.N. and Manley, T.R. 1986. Interrelationships between the concentrations of total condensed tannin, free condensed tannin and lignin in *Lotus sp.* and their possible consecuences in ruminant nutrition. *Journal of the Science of Food and Agriculture,* 37: 248-254.

Blum, A. and Ebercon, A. (1981) Genotype responses in sorghum to drought stress. III. Free proline accumulation and drought resistance. *Crop Sci.16*: 379- 386.

Boursier, P, and A.Lauchli.1990.Growth responses and mineral nutrient relatiion of salt -stressed Sorghum. Crop Sci. *30*: 1226-1233

Chaurasia, M., D.R. Chauhan and J.Singh, 2006. Effect of irrigation , nitrogen and phosphorus levels on fodder production of bajri (*Pennisetum glaucum L.*) –A local race of bajra. *Forage Res.* 32 (2): 128-129.

Dixon, R.A. and Pajva, N.L. 1995. Stress-induced phenylpropanoid metabolism. *The Plant Cell*, 7, 1085-1097.

Foo, L.Y., Newman, R., Waghorn, G., MCNABB, W.C. and Ulyatt M.J. 1996. Proanthocyanidins from *Lotus corniculatus. Phytochemistry*, 41: 617-624.

Dostanova, R.K., Klysheve, L.K. and Toibaeva, K.A. (1979). Phenol Compounds of Pea Roots Under Salinization of the Medium. *Fiziologiya-i- Biokhimiya Kul'turnykh Rastenii.* 11: 40-47.

Gebrehiwot, L., Beuselinck, P.R. and Roberts, C.A. 2002. Seasonal variations in condensed tannin concentration of three *Lotus* species. *Agronomy Journal,* 94: 1059-1065.

Ghassemi, F., Jakerman , A.J., and Nix,H.A. (1995). Salinisation of land and water resources: human causes, extent, management and case studies. CAB International, Wallingford, UK, p. 526.

Giorgini, J.P. and Suda, C.N.K .(1990). Ribonucleic acid synthesis in embryonic axes of coffee (*Coffea arabica* L. cv. Mundo Novo). *Revista Brasileira de Botânica.* 13: 1-9.

Ghoulam, C., Foursy, A. and Fares, K . (2002). Effect of salt stress on growth, inorganic ions and proline accumulation in relation of osmotic adjustment in five sugar beet cultivarts. *Environ. Exp. Bot.* 47: 39-50.

Gupta, K., D.S. Rana, and R.S. Sheoran, 2007. Response of forage sorghum to *Azospirillum* under organic and inorganic fertilizers. Forage Res.,33(3):168-170

Gupta , K ., Rana .D.S and Sheoran, R.S. (2008). Response of nitrogen and phosphorus levels on forage yield and quality of forage sorghum (*Sorghum bicolor (L.) Monch*). Forage Res.). Forage Research. 34 (3) :156-159

Gupta, K and Rana, D.S.(2009). Effect of integrated nutrient management on yield and quality of single cut forage sorghum . In " Emerging Trends in Forage Research and Live Stock Production", (2009) (Eds. S.K. Pahuja., U. N. Joshi, B.S.Jhorar, R.S. Sheoran) , Indian Society of Forage Research, CCS Haryana Agricultural University, Hisar, 125004, India, pp 158-159

Kamalak, A., O. Canibolat, Y. Gurbuz, O. Ozay and E. Ozkose, 2005. Chemical composition and its relationship to *in vitro* gas production of several tannin containing trees and shrub leaves. *Asian- Aust.J. Anim. Sci.,* 18 : 203-208.

Khan, M.A., I.A. Ungar, A.M. Showalter. (2000). Effect of sodium chloride treatments on growth and ion accumulation of halophyte *Haloxylon recurvum. Soil. Sci. Plant Anal.* 31: 2763-2774.

Khatri, R.S., Mohanraj, K., Gopalan, A., Shanmugnathan (1997). Genetic parameters for phydroorganic acid content in forage sorghum. *Sorghum bicolor* L. Moench. *J. Agric. Sci.* 2: 59-62.

Kumar, S., C.R. Rawat., K. Singh, and , N.P. Melkana, 2004. Effect of integrated nutrient management on growth, herbabe productivity and economics of fodder sorghum (*Sorghum bicolor* L. Moench). *Forage Res.* 30 (3):140-144.

Kumar, S., C.R. Rawat., Shiva Dhar, and , S. K. Rai, 2005. *Indian J. Agric. Sci.*, 75: 340-342

Latha, V. M., Satakopan, V. N., Jayasree, H., (1989). Salinity induced changes in phenol and ascorbic acid content in groundnut *Arachis hypogaea* leaves. *Current Science* 58(3): 151-152.

Majeed, A., Nisar, M.F. and Hussain, K. (2010). Effect of saline culture on the concentration of Na^+, K^+ and Cl^- in *Agrostis tolonifera. Curr. Res. J. Biol. Sci.* 2: 76-82.

Munns, R., Husain, S., Rivelli, A.R., James, R.A., Condon, A.G., Lindsay, M.P., Lagudah, E.S., Schachtman, D.P., Hare, R.A. (2002). Avenues for increasing salt tolerance of crops, and the role of physiologically based selection traits. *Plant Soil* 247: 93-105.

Niezen, J.H., Waghorn, T.S., Waghorn, C.G. and Charleston, W.A.G. (1995). Growth and gastrointestinal nematode parasitism in lambs grazing either Lucerne *(medicago sativa)* or sulla *(Hedysarum coronarium)* which contains condensed tannins *Journal of agricultural cience,* 125, 281-289.

Norton, B.W., 2003. The nutritive value of tree legumes. In: Forage tree Legumes in Tropical Agriculture, R.C. Gutteridge, and H.M. Shelton (Eds.).

Otero, M.J. and Hidalgo, L.G. 2004. Tannins condensados en especies forrajeras de clima templado: efectos sobre la productividad de rumiantes afectados por parasitosis gastrointestinales (una revisión) *Livestock Research for Rural Development,* 16: *466-469.*

Panckurst, C.E. and Johns, W.T. (1979). Effectiveness of *Lotus* root nodules. *Journal of Experimental Botany*, 30: 1095-1107.

Paolocci, F., Bovone, T.,Tosti, N., Arcioni, S. and Damiani, F. 2005. Light and an exogenous transcription factor qualitatively and quantitatively affect the biosynthetic pathway of condensed tannins in *Lotus corniculatus* leaves. *Journal of Experimental Botany*, 56: 1093-1103.

Reinoso, H., Sosa, L., Ramirez, L. and Luna, V. (2004). Salt-induced changes in the vegetative anatomy of *Prosopis strombulifera* (Leguminosae). *Canadian Journal of Botany* 82: 618-628.

Rhoades, J.D. and Loveday. (1990). Salinity in irrigated agriculture, In: Americans Society of Civil Engineers. *Irrigation of Agricultural crops.* (Eds.): B.A. Steward and D.R. Nilson. *Am. Soc. Agron. Mono. 30*: 1089-1142.

Sarifi, M., Ghorbanli, M. and Ebrahimzadeh, H. (2007). Improved growth of salinity stress after incubation with salt pre-treatment mycorhizal fungi. *Plant physiol.* 164: 1144-1151.

Sheoran, R.S., and D.S. Rana, 2005. Relative efficiency of vermicompost and farmyard manure integrated with inorganic fertilizers for sustainable productivity of forage sorghum (*Sorghum bicolor* L. Moench). *Acta Agronomica Hungarica* 53 (3): 303-308

Sheoran, R.S., and D.S. Rana, 2006: Relative efficiency of Azotobactor and nitrogen fertilizer in forage sorghum (Sorghum bicolor L.) under semi arid conditions. *Forage Res.* 32 (2): 65-68

Singh, A.K. (2004). The physiology of salt tolerance in four genotypes of chickpea during germination. *J. Agric. Sci. Technol.* 6: 87-93.

Singh, P and H.K. Sumeriya . 2005. Response of forage sorghum cultivars to different nitrogen levels under Udaipur condition in Rajasthan. *Forage Res.* 31: 51-54

Singh, S.P., Luthra, Y.P. and Lodhi, G. P. 1995. Assessment of quantitative and qualitative losses caused by stem borer in forage sorghum. *Forage Res.* 21: 109-116.

Turan, M.A., Katkat, V. and Taban, S. (2007). Variations in proline, chlorophyll and mineral elements contents of wheat plants grown under salinity stress. *J. Agron.* 6: 137-141.

Turan, M.A., Türkmen, N. and Taban, N. (2007). Effect of NaCl on stomatalresistance and proline, chlorophyll, Na, Cl and K concentrations of lentil plants. *J. Agron.* 6: 378-381.

Waghorn, C.G. and Shelton, I.D. 1997. Effect of condensed tannins in *Lotus corniculatus* on the nutritive value of pasture for sheep. *Journal of Agricultural Science,* 128: 365-372.

In: Sorghum: Cultivation, Varieties and Uses
Editors: Tomás D. Pereira

ISBN: 978-1-61209-688-9
©2011 Nova Science Publishers, Inc.

Chapter 5

DUAL-PURPOSE SORGHUM FOR GRAIN AND BIOENERGY PRODUCTION FOR SEMI-ARID AREAS OF SOUTHERN AFRICA

Itai Makanda[1,2], Pangirayi Tongoona[1] and John Derera[1]

[1]African Centre for Crop Improvement, School of Agricultural Sciences and Agribusiness, University of KwaZulu-Natal, P. Bag X01, Scottsville, Pietermaritzburg 3209, South Africa
[2]Agricultural Research Council (ARC)-Grains Crops Institute, P. Bag X1251, Potchefstroom 2520, South Africa

ABSTRACT

Sorghum (*Sorghum bicolour* L Moench) is one of the most important cereal crops in sub-Saharan Africa (SSA). It plays a major food security role in the semi-arid areas in southern Africa where small-scale and resource-poor farmers reside. Apart from food, sorghum has high potential as an industrial crop in the biofuel industry. Sorghum varieties have been traditionally developed for grain, especially in SSA, whereas specialist varieties for fodder or stem sugar have been bred in developed countries. What could be most appealing to small-scale farmers in SSA are dual-purpose varieties that combine high grain yield and stem sugar content. When developed, such varieties would be beneficial to these resource-poor farmers by providing adequate grain for food and sugar rich stalks for commercial production of bioethanol (used as biofuel) with possible multiplier effects. However, a survey of the literature indicates clearly that breeding effort has been limited to specialised cultivars alone. Information on the inheritance, combining ability, gene action and relationships between stem sugar and grain yield traits in breeding dual-purpose sorghum source germplasm is limited. This information is crucial for devising appropriate strategies for developing dual-purpose sorghum varieties. If grain yield and stem sugar are mutually exclusive, then breeding for the two traits in a single cultivar will be a challenging task. Results from a few studies have indicated a weak and non-significant association between these traits. These findings pertain to the populations and test environments which were sampled; therefore, a careful selection of base populations is key to achieving progress in breeding such cultivars in southern Africa. Further, there is also lack of information about perceptions and views of stakeholders on the potential of dual-purpose sorghum production, utilisation and the general value chain. Therefore, this review is an attempt at elucidating the information necessary to establish a viable breeding programme for dual-purpose sorghum cultivars. It covers the utilisation of sorghum for food and biofuel, benefits of the cultivars to the farmers, stakeholders' views on the technology; the genetics of grain yield and stem sugar content; and the relationship between the two traits. Implications of the findings and suggestions for the way forward are also discussed.

5.1. SORGHUM

Sorghum (*Sorghum bicolour* L. Moench) is a self-pollinating cereal crop thought to have originated in north-eastern Africa around Ethiopia, Sudan and East Africa (Dogget, 1988, de Wet and Harlan, 1971; Kimber, 2000; Acquaah, 2007). Some researchers argue for multiple centres of origin for the crop (Snowden, 1936; de Wet and Huckabay, 1967). Its distribution around the world is attributed to movement of people and its diversity to disruptive selection in different habitats, especially in northeast Africa. For example, it is argued that as *S. bicolor* moved west, it crossed with the wild *S. arundinaceum* giving rise to the type known as the durra sorghum (Kimber, 2000). There are four wild (*S. bicolor* subspecies *verticilliflorum*) and five cultivated (*S. bicolor* subspecies *bicolor*) races in sorghum, differentiated by head type, grain size, yield potential, and adaptation, among other traits (Acquaah, 2007). The cultivated races are bicolor, guinea, kafir, caudatum and durra (Kimber, 2000). In commercial sorghum breeding programmes, there are established working groups, namely kafir, milo, margaritiferum, feterita, hegari, shalu, kaoliang and zera-zera (Menz et al., 2004; Acquaah, 2007). Some intermediate races resulting from inter-mating are also recognised. The significance of the working groups is in differences in adaptation, yield potential and their implications to crop improvement. Researchers argue that the races are the best basis for grouping sorghum into heterotic groups for hybrid programmes (Menz et al., 2004) although most breeders use the A/B (male-sterile female group) and R (restorer male group) as the basis of heterotic grouping (Acquaah, 2007).

5.2. PRODUCTION AND UTILISATION

Sorghum total production ranks fifth among the important cereal crops worldwide and is second after maize in Africa (Chantereau and Nicou, 1994). Most of the sorghums planted, especially in Africa, are grain sorghum cultivars. The world produces about 63 Million tons of sorghum grain from about 46 Million ha of which 26 Million tons comes from Africa (from about 29 Million ha) and 157,000t is produced in southern Africa from an area of 220,000 ha (FAOSTAT, 2009). Apart from grain, sorghum can be used for sugar production using the sweet stem types that accumulate fermentable sugars ($\geq 8\%$) in their stems. These sweet sorghums provide an avenue for transforming sorghum into an industrial crop. Stem sugar can be used for bioethanol production and the grain for food. This dual-purpose nature of sorghum can provide rural households with dietary energy as well as the much needed income for other requirements from the small pieces of arid land from which they subsist. However, no effort has been made to combine grain and stem sugar in a single cultivar to produce a dual-purpose sorghum variety in southern Africa regardless of the presence of such varieties being reported in other regions (FAO, 2002; Reddy *et al.*, 2005). The potential for both grain yield, stem sugar accumulation and stem biomass yields in dual-purpose sorghum cultivars for the region has not been documented. This is evidenced by the non-availability of the dual-purpose sorghum cultivars on the market. Generating information on the behaviour of the traits after combination is important for a dual-purpose sorghum breeding programme. An ideal dual-purpose sorghum should achieve acceptable grain yields, stem brix and stem biomass. Although there are no set values of these traits, minimum grain yield of 1.5t ha^{-1}, stem brix of 11°brix and stem biomass of 30t ha^{-1} can be arbitrarily set to achieve about 3000l of bioethanol ha^{-1} and food security. Assuming a farmer plants two hectares, 3t of grain and 6000l of ethanol can be produced based on studies by Woods (2000) and Tsuchihashi and Goto (2004).

Figure 5.1. Types of specialised sorghum cultivars (a) grain sorghum and (b) sweet stem sorghum.

The potential for generating bioethanol from sweet sorghum has not been quantified in most countries and environments in southern Africa. Stem sugar values of between 10% and 25% (10°brix and 25°brix) have been reported in the literature (Woods, 2000; Reddy *et al.*, 2005; Tsuchihashi and Goto, 2004, 2008; Makanda *et al.*, 2009; Makanda, 2009). Stem sugar can be processed into jaggery or distilled to produce ethanol (FAO, 2002). Bioethanol yield of 3000l to 7000l ha^{-1} have been reported from biomass levels of between 30t to 120t ha^{-1} in Zimbabwe (Woods, 2001); Mozambique and South Africa (Makanda *et al.*, 2009), Romania (Roman *et al.*, 1998); Italy (Dolciotti *et al.*, 1998); United States of America (Anderson, 2005); China (FAO, 2002); and various European Union countries (Claassen *et al.*, 2004). Mean grain yield potentials of between 1.0t and 6.5t ha^{-1} have been reported with improved grain sorghums in Zambia, Zimbabwe and Botswana (Obilana *et al.*, 1997). Generally, the grain sorghums are short with a large harvest index (Figure 5.1a) while sweet sorghums are generally tall with a low harvest index (Figure 5.1b). There are scarce reports on yield of the traits when combined into a single dual-purpose cultivar and the potential for improving both in such cultivars. This gives scope for more research into the dual-purpose sorghum in the southern African region targeting the farmers in the semi-arid areas who cannot afford growing specialised sweet sorghums, but can benefit from the sale of the sweet stalks after harvesting the grain.

5.3. WHY DUAL-PURPOSE SORGHUM?

The potential use of sweet sorghum for bioethanol production has been demonstrated in southern Africa in Zimbabwe by Woods (2001). The author showed that sweet sorghum can be successfully incorporated into the sugarcane (*Saccharum officinarum* L.) processing system. Countries in southern Africa, for example Malawi, Mozambique, South Africa, Swaziland, Zambia and Zimbabwe, which have viable sugar industries based on sugarcane, the major sugar producing crop, can benefit from dual-purpose sorghums. These countries can maximise the exploitation of the sugar mills by feeding sweet sorghum during the sugarcane off-season. Sweet sorghum has the advantage of wide adaptation; growth under dryland conditions and rapid growth (Reddy and Sanjana, 2003) which make it complement rather than compete with sugarcane. Communities around processing mills can grow dual-purpose sorghums and transport them to the mills leading to sustainable rural development, enhanced renewable energy production, higher health standards through cleaner biofuels, and improved food security from the grain. Farmers in most parts of southern Africa already grow sorghum, for example, in Musikavanhu, a dryland communal area in Zimbabwe, Chivasa *et al.* (2001) reported that sorghum was grown by 94% of the farmers occupying 82% of the land. Derera *et al.* (2006) reported similar findings in the same area. Farmers already grow tall low grain yielding cultivars (Figure 5.2a) and the alternative dual-purpose cultivars are tall but higher grain yielding (Figure 5.2b). Farmers will therefore be changing the cultivar for a higher grain yielding one, which might mean faster adoption rate of dual-purpose sorghum.

Figure 5.2. (a) Farmers' current grain sorghum and (b) candidate dual-purpose sorghum cultivar under experimentation.

Many world bodies express reservations on the use of crops for bioenergy due to potential competition for land with food crops. This is thought to push food prices beyond the reach of many thereby compromising food security. The use of crops that complement rather than compete with food crops for resources is therefore advocated for. This creates a niche for dual-purpose sorghum where farmers can harvest the grain for food and sell the stalks for sugar extraction, reaping twice from the same crop and piece of land. According to the FAO (Gnansounou et al., 2005), dual-purpose sorghums can give a yearly gross margin of U.S. $1300 ha^{-1} compared to only U.S. $1054 ha^{-1} for grain sorghum reported by Hugos et al. (2009). With specialised sweet sorghum, farmers can earn between U.S. $40 to U.S.$ 97 ha^{-1} more than for grain sorghum (PSciJourn, 2010). Further, the cost of producing a litre of bioethanol from sweet sorghum in Zimbabwe was reported at U.S. $0.19 compared to the then global prices of ethanol of U.S. $0.30 to $0.35 a litre (Woods, 2000). The challenge that may arise is the development of appropriate technologies and markets for the stems. However, with the projected future fossil fuel shortages and the increasing global call for cleaner environments, the adoption of sweet sorghum as a raw material for the production of alternative fuel is likely to increase. This will potentially increase the demand for sweet stem sorghum, making the enterprise viable.

5.4. STAKEHOLDERS' VIEWS AND PERCEPTIONS

This section is based on a survey case study conducted by the authors in Zimbabwe and South Africa. Two studies were conducted to establish the awareness and perceptions of stakeholders on the potential of dual-purpose sorghums in Southern Africa. One survey was carried out in the semi-arid tropical lowlands in Zimbabwe under the conditions of small-scale and resource-poor farmers while the other, targeting sugar industries, plant breeders, engineers, political leaders, economists and extension workers, was conducted in South Africa and Zimbabwe. Full results of the survey are presented in Makanda (2009). In brief, results showed that all stake holders concurred that dual-purpose sorghum production was a viable enterprise that could alleviate poverty, enhance food security, create rural employment and boost rural development especially in the semi-arid areas in southern Africa. Farmers showed willingness to adopt the cultivars if they were made available and the other stakeholders pledged support for the endeavour.

5.4.1. A Dual-purpose Sorghum Ideotype and Possible Competing Crops

Farmers described their ideal dual-purpose sorghum as one with high yield, early to medium maturity with large white grains. There were some traits specific to farmers within certain areas, for example early maturity was preferred in Chivi and Chipinge North whereas in Chipinge South, intermediate maturing varieties which can escape late season drought were preferred. Chivi, Chipinge North and South are semi-arid rural districts in Zimbabwe in which the survey was conducted where sorghum plays an important food security role. This was attributed to differences in rainfall patterns between the areas. Chipinge South receives slightly more rains than the former two areas, although both areas are within the drought prone regions of the country. This indicated breeding challenges that could be encountered due to the differences in preferences among the farmers hence different breeding objectives.

Based on the survey, the crops that could possibly compete with dual-purpose stem sorghum production in the-semi arid areas were maize currently produced for food and cotton for household income. Results showed that household land holding hardly exceeded 2.0ha (of semi-arid land) and more than 50% of the land was allocated to cotton with maize in second place and sorghum third. Dual-purpose sorghum with its advantages can replace both crops because it provides for both food and income requirements, the role currently satisfied by the two crops, maize and cotton. This can be achieved through the development of superior varieties such as hybrids.

5.4.2. Challenges

Stakeholders identified challenges that must be addressed for the success of dual-purpose sorghum production and industrial utilisation. These could be categorized into challenges relating to the cultivars, technical challenges and economic challenges. These and the suggested solutions are summarised in Figure 5.3. This chapter concentrates on the crop challenges and the focus is on crop improvement for stem sugar, grain yield and stem biomass.

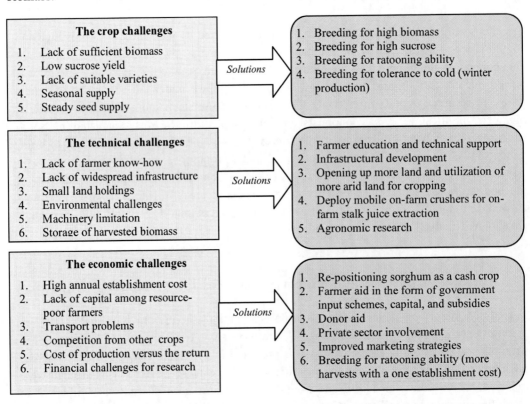

Figure 5.3. Challenges and possible solutions to the use of dual-purpose sorghum as a bio-energy crop in southern Africa.

5.5. PREREQUISITES FOR CULTIVAR DEVELOPMENT

5.5.1. The Need for Appropriate Breeding Source Germplasm

Cultivar development relies on the presence of genetic variability for the traits under consideration. It is, therefore, imperative as a first step to evaluate germplasm collections for grain yield potential, stem sugar accumulation and stem biomass potential to identify suitable parents. The next step is to understand the gene action controlling the traits. This information can be obtained by conducting combining ability studies that entails systematic crossing of selected parents using appropriate mating designs and subsequent experimental hybrid evaluation. Information on general combining ability (GCA) and specific combining ability (SCA) effects is critical in cultivar development. It determines the course of breeding, either through selection if GCA is important or inbreeding followed by hybridisation if SCA is important because variation due to GCA is attributed to additive genes and that due to SCA to non-additive gene action (Goyal and Kurmar, 1991; Cruz and Regazzi, 1994; Kenga et al., 2004). Lack of this information limits sorghum research under African dry-land conditions (Kenga et al., 2004). For dual-purpose sorghums, knowledge of the inheritance of the three major traits namely grain yield, stem sugar and stem biomass is important. A lot of work has been done on the inheritance of these traits separately, but not in dual-purpose germplasm.

5.5.2. Variation and Genetics of Grain Yield in Sorghum

Genetic variability in yield components is important for yield improvement programmes in sorghum (Warkad et al., 2008). Many traits in sorghum, especially those associated with inflorescence, largely determine yield levels of the different races described in earlier sections. Farmers are aware of this fact, for example, Abdi et al. (2002) reported that sorghum farmers in north-eastern coastal regions of Africa prefer the durra (compact head) types due to their high grain yields and quality. Head type is chiefly determined by the rachis and branch lengths, distance separating the whorls, and the angle of the branches from the rachis (Dogget, 1988). There are semi compact elliptic, compact elliptic, semi loose primary branches, very loose primary branches, very loose drooping primary branches, and half broomcorn head types in sorghum (Dogget, 1988; Abdi et al., 2002). These bring variation for grain yield potential. The guinea sorghums with less compact head types have been described as lowland sorghums and are generally low yielding compared to the durra types adapted to the high rainfall highlands (Dogget, 1988). Variations in yield also exist within each type which can be exploited in cultivar development. Variation in the grain yield and its components namely days to maturity, days to 50% flowering, plant height, number of leaves, head length and width and weight of 1,000 seeds have been reported within each type. Many African countries have a rich collection of sorghum germplasm. Such collections are important because the more diverse the genetic base, the more distant the lines developed and consequently the higher the hybrid vigour for yield and other important traits that can be realised and maintained on crossing (Li and Li, 1998). Genetic variability is therefore key to cultivar development.

In self-pollinating crops, advances in grain yield improvement have been realised through selection alone. Hybrids were confined to cross-pollinating crops such as maize. In recent developments, hybridisation has been successfully used to develop hybrid cultivars in self-pollinating crops like rice and sorghum (Li and Li, 1998; Kenga et al., 2004). This has enabled the exploitation of both gene additivity and non-additivity with combining abilities and heritabilities computations becoming common and important in sorghum breeding.

Grain yield has long been established to be a quantitatively inherited trait in many crops. Studies have shown both GCA and SCA effects to be important in many sorghum traits including grain yield (Haussmann et al., 1999; Tadesse et al., 2008; Kenga et al., 2004; Yu and Tuinstra, 2001; Makanda et al., 2010). Further, heritability in the narrow sense (h^2) of between 10 - 86% for grain yield and 91-99% for days to flowering have been reported (Warkad et al., 2008; Bello et al., 2007; Biswas et al., 2001; Lothrop et al., 1985; Haussmann et al., 1999). The significance of these findings is that both additive and non-additive gene effects are important for these traits. Therefore both selection and hybridisation improve grain yield. Parental line development through selection followed by hybrid production using parents with high GCAs and SCAs can therefore improve grain yield. Regardless of the availability of this information, it is important to note that combining ability and inheritance information cannot be generally applied as they are pertinent to a specific set of germplasm and environment (Falconer and Mackey, 1996). Further, the studies quoted above were conducted on specialised grain cultivars, not dual-purpose sorghums, making it important to study the gene action controlling the traits in dual-purpose sorghums to aid cultivar development. Studies by the authors showed both GCA and SCA to be important for grain yield in dual-purpose sorghum (Makanda, 2009). More studies are therefore necessary to ascertain these.

5.5.3. Genetics of Stem Sugar and Biomass

There is an increase in studies on the inheritance of stem sugar in sorghum. Earlier reports suggested partial dominance with Schlehuber (1945) reporting hybrids intermediate between the two parents in total solids and sucrose. Later on, a single gene "X" was reported to control sugar accumulation (Baocheng et al., 1986). However, new evidence suggests sugar accumulation to be under the control of recessive genes acting in an additive manner with broad sense heritability (H^2) estimates ranging from 0.65 to 0.81 in different populations (Guiying et al., 2000). These reports are consistent with studies by Schlehuber's (1945). Further studies found both gene additivity and non-additivity to be important for stem sugar with Makanda et al. (2009) reporting signifince in both GCA and SCA effects for the traits. Baocheng et al. (1986) reported GCA effects to be more important (10-26 times) than SCA effects and narrow sense heritabilities (h^2) between 0.40 and 0.96. Crosses between sweet and non-sweet sorghums produced transgressive segregants in the F_2 for stem sugar content in sweet sorghum (Guiying et al., 2000). This observation suggest that stem sugar in sorghum is under quantitative inheritance. Crosses between low sugar types resulted in negative heterosis for stem sugar accumulation, further giving evidence for gene additivity for stem sugar inheritance (Guiying et al., 2000). In recent QTL analyses (in sweet sorghum) by Natoli et al. (2002), no significant segregation for genes with major effects on sugar percentage were found but Ritter et al. (2008) reported QTL alleles from some entries to increasing sucrose content, sugar content and °brix. The moderate heritability estimates, significant GCA and SCA effects and reports of QTL controlling stem sugar suggest that the trait can be improved through traditional breeding approaches by applying selection and hybridisation.

Plant height is positively associated with plant biomass and therefore breeding tall dual-purpose cultivars translates to high biomass yields and subsequently high sugar yield per unit area. Plant height is known to be controlled by four dwarfing genes, Dw_1, Dw_2, Dw_3 and Dw_4, with tall being incompletely dominant to short (Rooney, 2000). The height differences are brought about through the control of internode length with plants with zero dwarfing genes ($Dw_1Dw_2Dw_3Dw_4$) growing up to 4.00m while those with the four recessive dwarfing genes ($dw_1dw_2dw_3dw_4$) being less than 0.50m (Rooney, 2000; Acquaah, 2007). Single ($Dw_1Dw_2Dw_3dw_4$), double ($Dw_1Dw_2dw_3dw_4$) and triple ($Dw_1dw_2dw_3dw_4$) recessive dwarfs average around 1.20-2.07m, 0.82-1.26m, and 0.52-0.61m, respectively (Rooney, 2000; Acquaah, 2007). Some interaction effects have also been reported to influence plant biomass accumulation in photosensitive cultivars. Rooney and Aydin (1999) reported duplicate recessive epistasis for plant height. However, the applicability of these findings for stem sugar and biomass on dual-purpose sorghum needs investigation because these reports were based on specialised sweet-stem sorghums.

5.5.4. Stem Sugar Distribution in Sorghum Stalks

Stem sugar variations in sweet sorghum have been reported along and across the stem as well as in time during the plant's growth cycle. Rose and Botha (2000) reported a sharp gradient in sugar increases between internode three and six in sorghum stems and slower increase between internode six and nine. This seems to imply that sucrose content of the core bottom tissue is lower compared to the upper tissues along the stem. This can be because the bottom tissue is concerned with non-sucrose storing metabolism, that is, metabolic processes involving respiration and growth hence more assimilates are committed and transported to the younger actively growing upper tissue (Rose and Botha, 2000).

Across the stem, the bark contains less sugars (glucose and sucrose) compared to the inner pith (Billa et al., 1997). As a percentage of dry weight, Billa et al. (1997) reported the bark to contain 32.2% sucrose and 2.4% glucose whereas the pith contained 67.4% and 3.7%, respectively. However, the bark had more lignin, cellulose and hemicelluloses compared to the pith. Ferraris and Charles-Edwards (1986) found stem sugar concentration to be a function of growth duration. They reported higher sugar concentration in one of their cultivars after grain initiation, particularly sucrose concentration. However, they reported low grain yields in other cultivars although remobilization of stem sugars to the grain was negligible. Zhao et al. (2009) reported ethanol production from stalk sugars to increase with increases in time after anthesis. This demonstrated an increase in stem sugar concentration during the grain-filling period, suggesting that little to no photo-assimilate translocation to the developing grain at the expense of the stems was absent. This suggests that dual-purpose sorghum cultivars can be developed.

In developing dual-purpose sorghums, remobilisation of photo-assimilates from the stem to the grain is not desirable because the ideal cultivar must retain high stem sugar at the same time achieve high grain yield. The possible complication that can arise is that lack of remobilisation might result in lower grain yield. The fore-mentioned studies showed that it is possible to identify genotypes with negligible remobilisation of photo-assimilates from the stem to the grain. Therefore, in dual-purpose cultivar development, it is necessary to screen materials to identify genotypes that combine both negligible or non-remobilisation of stem photo-assimilates and at the same time attain high grain yield. These can be used as source germplasm in a breeding programme for dual-purpose sorghums.

Overall, it has been shown that there is variation in sugars along the sweet sorghum stem and therefore, in the absence of whole stem crushers, taking sugar measurement at different points along the stem gives a better representation of the sugar performance compared to taking single measurements. The temporal variation necessitates the need for sampling sorghum stalks at the correct time during the growth cycle, and in dual-purpose sorghum, the sampling time must be the harvestable maturity stage of the grain because it is a major component.

5.5.5. Screening for Stem Sugar

There are many techniques used to quantify soluble sugars in plant tissues. These methodologies, which are well developed, are described by many researchers including Chow and Landhäusser (2004) and Reed et al. (2004). Most of these techniques follow common steps of drying plant sample, grinding it into powder, homogenising a sub-sample of the powder in a carbohydrate solvent (Hendrix, 1993), and then using various chemicals and steps to determine the amount of sugars in plants. The commonly used quantification methods when high levels of specificity and discrimination between the sugars is required include the high performance liquid chromatography (HPLC), calorimetry using anthrone reagent (Chow and Landhäusser, 2004; Reed et al., 2004), gas chromatography (GC), liquid chromatography (LC), and the enzyme methods based on NADPH absorption (Chow and Landhäusser, 2004). Each of these has its advantages and disadvantages ranging from cost, hazardous chemicals, low throughput, reagent stabilities, digestion of target compounds and others. The disadvantages common to all the methods are the low throughput, the need for sophisticated equipment, lack of adaptation to field conditions where the screening takes place and their being expensive.

Researchers are now advocating for more specific, time saving, and less hazardous methods (Hendrix, 1993). With large samples to be screened in the field, the refractometer is the most widely used instrument for stem sugar quantification. It measures the sugars in degrees brix (°brix), which is a quantification of soluble sugar-to-water mass ratio of fluid (Wikipedia, 2009). For example, if a 100g solution is 15°Bx, it means that 15 grams of the 100g is made up of the sugars. This scale is used to approximate the amount of sugars in many plant juice extracts incuding fruits, vegetables, sugarcane and sweet sorghum or in beverages like wine and soft drinks. The refractometer offers a cheap, time saving, and less hazardous method with high throughput. It can be used for screening a lager sample in the field. Recent improvements have seen the development of portable, simple and easy to use digital refractometers with automatic temperature compensation (ATC). An example of such portable refractometers is presented in Figure 5.4, an Atago -1 digital hand-held pocket refractometer with ATC ranging from 0.0 to 50°C. Although the refractometer has the disadvantage of inability to sample the whole stalk juice reading in the field as it depends on sectional cuttings, this can be addressed by measuring brix readings at different points along the length of the stalks so as to sample variation along the stem including nodes. Otherwise the whole stalks are crused, the juice extracted, mixed and then measured.

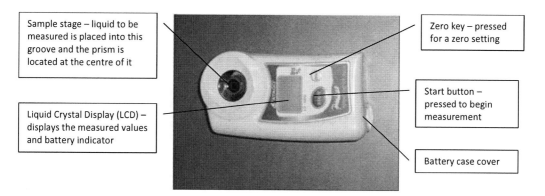

Figure 5.4. The Atago PAL-1 digital hand-held pocket refractometer used to measure stem sugar content (°brix) in the field.

5.6. Information that is Lacking

No work has been reported on the traits combined in dual-purpose sorghums. In the long run, the development of parental lines (base population) with high performance across the three traits is key to the dual-purpose sorghum cultivar breeding programme. These can be used as parents in hybrid cultivar development programmes or as pure line varieties where high performance is demonstrated. Makanda *et al.* (2009, 2010) studied the gene action controlling grain yield and stem sugar content in dual-purpose experimental hybrids and found both GCA and SCA to be significant for both traits. This implied that both traits were under the control of genes with both additive and interactive effects, suggesting quantitative inheritance for both traits (in combination) in a single cultivar. More studies are necessary to fully elucidate inheritance of these traits in dual-purpose sorghum cultivars.

Further, success in breeding dual-purpose sorghum for grain and stem sugar depends on the understanding of the relationship between the traits. The general notion is that improving grain yield results in a reduction in stem sugar yields. The argument is that the two represents two powerful sinks for the limited photo-assimilates. Based on this perception, it is assumed that high grain yielding cultivars are low in stem sugar and vice versa. However, there is no evidence to support this view. Reports on the relationship are scarce. Reports by Guiying *et al.* (2000) suggested a negative relationship (r = -0.472) between stem sugar and weight of 1,000 seeds, a grain yield components. Conclusions based on studies including 1,000 seed weight alone can be erroneous because grain yield is dependent on the number and size of the seeds per plant. Therefore, using grain yield represents the most dependable results on the relationship. This necessitates the detailed studies on the relationship between grain yield components and stem sugar traits if acceptable dual-purpose sorghum cultivars are to be developed. Makanda (2009) performed a correlation and path-coefficient analysis of grain yield components and stem sugar traits and found that for the top performing dual-purpose type experimental hybrids and lines, there was a weak, positive and non-significant relationship (r = 0.1470). For all entries used, which included specialised grain and sweet stem sorghum, the relationship was very weak, positive and also non-significant (r = 0.071). This implied that the general notion of a negative relationship might not be true or cannot be generally applied. Therefore the choice of source germplasm might be of paramount importance in developing dual-purpose sorghums. Apart from this work, the authors found no other work dealing with this relationship although an understanding of this relationship helps breeders formulate and optimize the selection strategy.

5.7. ENHANCING PRODUCTION

5.7.1. Hybrid Cultivar Development

Sorghum breeding history is not as long and successful as that of other grain cereals such as maize, wheat, rice and barley. However, the past years have seen significant advances with the advent of hybrid seed production as a result of the discovery and use of the male sterility system. Using hybrid sorghum, India has achieved an 80% yield increase in sorghum in the last 20 years on a 37% background decline in area under the crop (Kenga et al., 2004). The authors also reported that in the same period, Africa's area under sorghum had nearly doubled but yields have not increased. Apart from genetic research, the success story of Indian sorghum has been attributed to hybrid use (Kenga et al., 2004). Elsewhere, such as in China, production of hybrid varieties has become predominant. Consequently sorghum breeding programmes are emphasising hybrids, although traditional population improvement procedures are still in use (Li and Li, 1998). About 90% of China's sorghum growing land is under hybrid cultivars which have led to several fold yield increases in China (Li and Li, 1998). Haussmann et al. (1999) concluded that hybrid production had the potential to increase yield in semi-arid areas of Kenya, a sentiment that has been demonstrated in the generality of southern Africa.

5.7.2. Exploitation of the Off-season

Successful production of dual-purpose sorghum might entail expansion of production to include off-season production in tropical lowlands that have optimum temperatures and water supply throughout the year. These areas include the Zambezi basins covering parts of Zimbabwe, Mozambique, Zambia and Malawi, the Makhathini flats in South Africa and Chokwe in Mozambique. Off-season production in Chokwe and Makhathini flats was demonstrated to give optimum grain yield, stem biomass and stem sugar percentage with dual-purpose sorghum experimental hybrids (Makanda et al., 2009; 2010). Off-season production for grain was reported to significantly contribute to food security in Somalia (Food Security Assessment Unit, 1998) and India (Patil, 2007), while all year round production of sweet sorghum was demonstrated in Indonesia (Tsuchihashi and Goto, 2008). There is potential to use the off-season, in addition to the traditionally used in-season production environments for dual purpose sorghum. The limitation is the non-availability of appropriate dual-purpose sorghum cultivars for this purpose and cultivar development is viewed as key. Therefore, it is important to generate the information required towards the devising of an appropriate strategy towards the development of dual-purpose sorghum cultivars that are also adapted to the off-season production environments.

5.8. RECOMMENDATIONS FOR FUTURE STUDIES

The recommendation for future studies towards the development and successful use of dual-purpose sorghums in southern Africa can be summarised in the following points:

- There is need for extensive studies for a full elucidation of the genetics and inheritance of stem sugar, biomass and grain yield and the relationship of the traits in dual-purpose sorghums. This entails the development of dual-purpose base germplasm from which dual-purpose sorghum cultivars can be developed, be they pure line cultivars or hybrid cultivars, the latter of which have been conclusively proved to be more productive compared to the former.
- Hybrid development is dependent on the availability of male (restorer (R) lines), male-sterile female lines (A lines) and the maintainer lines (B lines) to the A lines. Currently, southern Africa lacks locally developed male-sterile lines for hybrid production. Apart from the need for male sterile-lines in general, particular focus on dual-purpose male-sterile lines is required for the successful development of a dual-purpose sorghum hybrid breeding programme.
- Further, economic, technical and feasibility studies on the technology in different countries in the region are of paramount importance.
- Lastly, there is need for considerable capital investments to enable small-scale farmers to successfully adopt and use the technology, as put forward in the survey of the stakeholders.

REFERENCES

Abdi, A., Bekele, E., Asfaw, Z. and Teshome, A. 2002. Patterns of morphological variation of sorghum (*Sorghum bicolor* L. Moench) landraces in qualitative characters in North Shewa and South Welo, Ethiopia. *Hereditas* 137: 161-172.

Acquaah, G. 2007. *Principles of plant genetics and breeding.* First edition, Blackwell, Malden, U.S.A., pp 509-518.

Anderson, I.C. 2005. *Ethanol form sweet sorghum.* Iowa Energy Centre, Iowa State University. http://www.energy.iastate.edu/renewable/biomass/cs-anerobic2.html. Accessed 09 May 2006.

Bello, D., Kadams, A.M., Simon, S.Y. and Mashi, D.S. 2007. Studies on genetic variability in cultivated sorghum (Sorghum bicolor L. Moench) cultivars of Adamawa State Nigeria. *American-Eurasian Journal of Agriculture and Environmental Sciences* 2: 297-302.

Billa, E., Koullas, D.P., Monties, B. and Koukios, E.G. 1997. Structure and composition of sweet sorghum stalk components. *Industrial Crops and Products* 6: 297-302.

Biswas, B.K., Hasanuzzaman, M., Eltaj, F., Alam, M.S. and Amin, M.R. 2001. Simultaneous selection for fodder and grain yield in sorghum. *Journal of Biological Sciences* 1: 319-320.

Chantereau, J. and Nicou, R. 1994. *Sorghum. The Tropical Agriculturist Series.* CTA Wageningen, Netherlands; Macmillan, London.

Chivasa, W., Chiduza, C., Nyamudeza, P. and Mashingaidze, A.B. 2001. Biodiversity on-farm in semi-arid agriculture: A case study from a smallholder farming system in Zimbabwe. *The Zimbabwe Science News* 34: 13-18.

Chow, P.S. and Landhäusser, S.M. 2004. A method for routine measurements of total sugar and starch content in woody plant tissues. *Tree Physiology* 24: 1129-1136.

Claassen, P.A.M., de Vrije, T., Budde, M.A.W., Koukios, E.G., Glynos, A. and Réczey, K. 2004. Biological hydrogen production from sweet sorghum by thermophilic bacteria. *2nd World conference on biomass for energy, industry and climate protection,* 10-14 May 2004, Rome, Italy, pp 1522 – 1525.

Cruz, C.D. and Regazzi, A.J. 1994. *Modelos biométricos aplicados ao melhoramento genético.* Viçosa, Imprensa Universitária, 394p. (English translation).

de Wet, J.M.J. and Huckabay, J.R. 1967. The origins of *Sorghum bicolor*. II. Distribution and domestication. *Evolution* 211: 787-802.

de Wet, J.M. and Harlan, J.R. 1971. The origins and domestication of sorghum bicolor. *Economic Botany* 25: 128-135.

Dogget, H.D. 1988. *Sorghum.* 2nd, Edition. Longman Group UK, Green and Co. Ltd: London, Harlow, England, 512pp.

Dolciotti, I., Mambelli, S., Grandi, S. and Venturi, G. 1998. Comparison of two sorghum genotypes for sugar and fibre production. *Industrial Crop Products* 7: 265-272.

Falconer, D.S. and Mackay, T.F.C. 1996. *Introduction to quantitative genetics,* 4rd edition. Longman, Essex, England.

FAO, 2002. Sweet sorghum in China. Agriculture 21 Magazine http://www.fao.org/ag/magazine/0202sp2.htm. Accessed 09 May 2006.

Ferraris, R. and Charles-Edwards, D.A. 1986. A Comparative analysis of growth of sweet and forage sorghum crops. II accumulation of soluble carbohydrates and nitrogen. *Australian Journal of Agricultural Research* 37: 513-522.

Food Security Assessment Unit. 1998. *Off-season production survey in southern Somalia January-March 1998* http://www.fsausomali.org/uploads/Other/274.pdf Accessed 16 February 2009.

Gnansounou, E., Dauriat, A. and Wyman, C.E. 2005. Refining sweet sorghum to ethanol and sugar. Economic trade-offs in the context of North China. *Bioresource Technology* 96: 985-1002.

Goyal, S.N. and Kumar, S. 1991. Combining ability for yield components and oil content in sesame. *Indian Journal of Genetics and Plant Breeding* 51: 311–314.

Guiying, L., Weibin, G., Hicks, A. and Chapman, K.R. 2000. A training manual for sweet sorghum. FAO/CAAS/CAS, Bangkok, Thailand. http://ecoport.org/ep?SearchType=earticleView&earticleId=172&page=2273 Accessed 15 November 2007.

Hugos, F., Makombe, G. Namara, R.E. and Awulachew, S.B. 2009. *Importance of irrigated agriculture to the Ethiopian economy: capturing the direct benefits of irrigation.* Colombo, Sri Lanka: International Water Management Institute. 37p. (IWMI Research Report).

Haussmann, B.I.G., Obilana, A.B., Ayiecho, P.O., Blum, A., Schipprack, W. and Geiger, H.H. 1999. Quantitative-genetic parameters of sorghum (*Sorghum bicolour* L. Moench) grown in Semi-Arid Areas of Kenya. *Euphytica* 105: 109-118.

Hendrix, D.L. 1993. Rapid extraction and analysis of non-structural carbohydrates in plant tissues. *Crop Science* 33: 1306-1311.

Kenga, R., Alabi, S.O. and Gupta, S.C. 2004. Combining ability studies in tropical sorghum (*Sorghum bicolor* L. Moench). *Field Crops Research* 88: 251-260.

Kimber, C.T. 2000. Early domestication of sorghum and its early diffusion to India and China. In: Smith, C.Y. and Fredericksen, R.A. (eds.), 2000. *Sorghum, History, Technology, and Production*. John Wiley & Sons, Inc., pp 3-98.

Li, Y. and Li, C. 1998. Genetic contribution of Chinese landraces to the development of sorghum hybrids. *Euphyitica* 102: 47-55.

Liang, G.H. and Walter, T.L. 1968. Heritability estimates and gene effect for agronomic traits in sorghum, Sorghum vulgare *Pers. Crop Science* 8: 77–83.

Lothrop, J.E., Atkins, R.E. and Smith, O.S. 1985. Variability for yield and yield components in IAAR grain sorghum random mating populations. I: Means, variance components, and heritabilities. *Crop Science* 8: 235-240.

Makanda I. 2009. *Combining ability and heterosis for stem sugar traits and grain yield components in dual-purpose sorghum (Sorghum bicolor L. Moench) germplasm.* PhD Thesis, Faculty of Science and Agriculture, University of KwaZulu-Natal, South Africa.

Makanda I., Tongoona, P., Derera, J., Sibiya, J. and Fato, P. 2010. Combining ability and cultivar superiority of sorghum germplasm for grain yield across tropical low and mid altitude environments. *Field Crops Research* 116: 75 – 85.

Makanda, I., Tongoona, P. and Derera, J. 2009. Combining ability and heterosis of sorghum germplasm for stem sugar traits under off-season conditions in tropical lowland environments. *Field Crops Research* 114: 272-279.

Menz, M.A., Klein, R.R., Unruh, N.C., Rooney, W.L., Klein, P.E. and Mullet, J.E. 2004. Genetic diversity of public inbreds of sorghum determined by mapped AFLP and SSR markers. *Crop Science* 44: 1236-1244.

Natoli, A., Gorni C., Chegdani, F., Marsan, P.A., Colombi, C., Lorenzoni, C. and Marocco, A. 2002. Identification of QTLs associated with sweet sorghum quality. *Maydica* 47: 311-322.

Patil, S.L. 2007. Performance of sorghum varieties and hybrids during postrainy season under drought situations in Vertisols in Bellary, India. *Journal of SAT Agricultural Research:* 5(1).

PSciJourn, 2010. *Biofuels has to be science-based.* The Philippine Science Journalists Association Inc http://pscijourn.wordpress.com/, Accessed 10 April 2010.

Reddy, B.V.S. and Sanjana, S.P. 2003. Sweet sorghum: characteristic and potential. *International Sorghum and Millets Newsletter* 44: 26-28.

Reddy, V.S.B., Ramesh, S., Reddy, P.S., Ramaiah, B., Salimath, P.M. and Kachapur, R. 2005. Sweet sorghum – a potential alternate raw material for bio-ethanol and bioenergy. *International Sorghum and Millets Newsletter* 46: 79–86.

Reed, A.B., O'Connor, C.J., Melton, L.D. and Smith. B.G. 2004. Determination of sugar composition in grapevine rootstock cuttings used for propagation. *American Journal of Enology and Viticulture* 55: 181-184.

Ritter, K.B., Jordan, D.R., Chapman, S.C., Godwin, I.D., Mace, E.S. and McIntyre, C.L. 2008. Identification of QTL for sugar-related traits in a sweet × grain sorghum (*Sorghum bicolor* L. Moench) recombinant inbred population. *Molecular Breeding* 22: 367–384.

Roman, Gh.V., Hall, D.O., Gosse, Gh., Roman, A.N., Ion, V. and Alexe, G.H. 1998. Research on sweet-sorghum productivity in the South Romanian plain. *Agricultural Information Technology in Asia and Oceania* 183-188.

Rooney, W.L. 2000. Genetics and Cytogenetics. In Smith, C.W. and Frederiksen, R.A. (eds.) *Sorghum origin, history, technology, and production.* John Wiley and Sons, Inc. pp 261 - 308.

Rooney, W.L. and Aydin, S. 1999. Genetic control of a photoperiod-sensitive response in *Sorghum bicolor* (L.) Moench. *Crop Science* 39: 397-400.

Rose, S. and Botha, F.C. 2000. Distribution patterns of neutral in-vertase and sugar content in sugarcane internodal tissue. *Plant Physiology and Biochemistry* 38: 819-824.

Schlehuber, A.M. 1945. Inheritance of stem characters: In certain sorghum varieties and their hybrids. *Journal of Heredity* 36: 219–222.

Snowden, J.D. 1936. *The cultivated races of sorghum.* Adlard and Sons, London.

Tadesse, T., Tesso, T. and Ejeta, G. 2008. Combining ability of introduced sorghum parental lines for major morpho-agronomic traits. *SAT eJournal (An open Access Journal published by ICRISAT)* 6: 1-7.

Tsuchihashi, N. and Goto, Y. 2004. Cultivation of sweet sorghum (*Sorghum bicolor* (L.) Moench) and determination of its harvest time to make use as the raw material for fermentation, practiced during rainy season in dry land Indonesia. *Plant Production Science* 7: 442-448.

Tsuchihashi, N. and Goto, Y. 2008. Year-round cultivation of sweet sorghum [*Sorghum bicolor* (L.) Moench] through a combination of seed and ratoon cropping in Indonesian Savanna. *Plant Production Science* 11: 377-384.

Warkard, Y.N., Potdukhe, N.R., Dethe, A.M., Kahate, P.A. and Kotgire, R.R. 2008. Genetic variability, heritability and genetic advance for quantitative traits in sorghum germplasm. *Agricultural Science Digest* 28: 165-169.

Wikipedia, *The free encyclopaedia,* 2009. Brix. http://en.wikipedia.org/wiki/Brix Accessed 10 June 2009.

Woods, J. 2000. *Integrating sweet sorghum and sugarcane for bioenergy: modelling the potential of electricity and ethanol production in SE Zimbabwe,* PhD thesis, Kings College London.

Woods, J. 2001. The potential for energy production using sweet sorghum in southern Africa. *Energy for Sustainable Development* V: 31-38.

Yu, J. and Tuinstra, M.R. 2001. Genetic analysis of seedling growth under cold temperature stress in grain sorghum. *Crop Science* 41: 1438-1443.

Zhao, Y.L., Dolat, A., Steinberger, Y., Wang, X., Osman, A. and Xie, G.H. 2009. Biomass yield and changes in chemical composition of sweet sorghum cultivars for biofuel. *Field Crops Research* 111: 55-64.

In: Sorghum: Cultivation, Varieties and Uses
Editors: Tomás D. Pereira

ISBN: 978-1-61209-688-9
©2011 Nova Science Publishers, Inc.

Chapter 6

TOLERANCE TO ALUMINUM AND NUTRIENT STRESS IN SORGHUM GROWN UNDER SIMULATED CONDITIONS OF TROPICAL ACID SOILS

M. Shahadat Hossain Khan[1], Afrin Akhter[1], Tadao Wagatsuma[2], Hiroaki Egashira[2], Idupulapati M. Rao[3], S. Ishikawa[4]

[1]HMD Science and Technology University, Dinajpur-5200, Bangladesh
[2]Faculty of Agriculture, Yamagata University, Yamagata 997-8555, Japan
[3]Centro Internacional de Agricultura Tropical, A. A. 6713, Cali, Colombia
[4]National Institutes for Agro-Environmental Sciences, Ibaraki, 305-8604, Japan

ABSTRACT

Acid soils occur mainly in two global belts: the northern belt, with cold and humid conditions, and the southern tropical belt, with warmer humid conditions. Although aluminum (Al) is usually regarded as the determining factor for growth of many crop plants in acid soils, nutrient deficiencies are also a major predicament in tropical acid soils. Sorghum (*Sorghum bicolor* Moench [L]) is generally an Al-sensitive crop species. While screening for the differential Al tolerance among sorghum cultivars, we found that lower level of Al in culture solutions increased the coefficient of variation in sorghum cultivas than in the case of maize cultivars. Contrary to the common agreement on the greater contribution of Al tolerance for improved adaptation to acid soils, appropriate strategy is needed for sorghum production in tropical acid soils. This observation is based on the comprehensive but preliminary evaluation under long-term conditions of combined stress conditions with varying concentrations of Al and nutrients in solution culture simulating the nutrient stress of tropical acid soils. Limited research has been carried out considering these two factors, i. e., concentrations of Al and nutrients, concurrently. In sorghum, a greater tolerance under combined stress conditions was associated with a higher shoot K concentration. Although Al tolerance is considered as an important strategy for sorghum production in tropical acid soils, plant nutritional characteristics linked to low nutrient tolerance can be the primary factor for better growth of maize cultivars that are tolerant to Al.

INTRODUCTION

Approximately thirty per cent of the world's ice free land areas are acidic in nature. According to Sanchez and Salinas (1981) approximately 55% of the soils in America, 39% in tropical Africa and 37% in Tropical Asia are acidic representing 1.6 billion hectares. Acid soils occur mainly in two global belts: the northern belt with cold and humid temperate climate and the southern tropical belt, with warmer and humid conditions. In southern belt, highly acidic soils are found mainly in south America- for example Peru, Brazil, Colombia, in Mid Africa, for example, Congo, Central African Republic and in some countries in South East Asia. Within Asia, strongly weathered acid soils mainly located in Bangladesh, Thailand, Cambodia, Laos, Indonesia and Vietnam. Acidification of southern belt mainly occurs due to natural phenomenon i.e. leaching of basic cations due to high weathering as a result of high temperature and heavy rainfall within a short rainy season. Naturally occurring acid soils mainly falls into 4 soil orders. These are Oxisols (Ferralsols), Ultisols (Acrisols, Nitosols, Planosols), Andisols (Andosols), and Alfisols (Podzoluvisols) (Sumner and Noble 2003). The major problems of acid soils in these regions (e.g. South American Savannas) are the low content of cations, the toxicity of exchangeable Al and/or soluble Al and low levels of phosphorus and low silicon availability as a result of extensive weathering (Okada and Fischer 2001, Rao 2001, Rao et al. 1993).

Tropical soils may be defined as all the soils of the geographic tropics that are in the region of the earth lying between the tropic of Cancer and the tropic of Capricorn (Eswaran et al. 1992). There are many causes for the poor growth of plants in these soils. The common and primary stress factors are: 1) H^+ toxicity/low pH, 2) Al and Mn toxicities and 3) a deficiency of essential nutrients (N, P, K, Ca, Mg, Mo and B) (Rao 2001; Rao et al. 1993). A major problem of these soils is the low content of cations, the toxicity of exchangeable Al and/or soluble Al and low levels of phosphorus as well as low silicon availability due to long weathering (Okada and Fischer 2001).

Numerous studies have been focused on the identification of major factors that cause a decrease in plant production in solution culture with or without Al and its related mechanisms (Ofei-Manu et al. 2001, Pavan et al. 1982, Pintro and Taylor, 2004, Wright et al. 1989). However, the number of studies that have simultaneously considered the two major stress factors (high Al and low-nutrients) in tropical acid sols is limited (Wenzl et al. 2003). In high-nutrient solution, Al toxicity is alleviated by a physicochemical interaction between Al and other ions, the formation of non-toxic complexes with OH^- and SO_4^{2-}, and the precipitation of Al phosphates from high ionic strength solutions (Blamey et al. 1983, 1991, Wheeler and Edmeads 1995). The ionic strength of tropical savanna soil solutions varies from 1.3 to 1.7 mmol L^{-1} for unfertilized samples and increases to 5.4 – 13.4 mmol L^{-1} after fertilization (Wenzl et al. 2003) which is far lower than that of average mineral soils. Wenzl et al. (2001) reported an inadequate supply of nutrients may be one of the main factors that contribute to the poor persistence of Al sensitive *Brachiaria ruziziensis* in infertile acid sols. On the other hand, Al activities of a solution of tropical acid soil ranged from 2.26 to 196.5 µmol L^{-1}

(Pintro et al. 1999). Blamey et al. (1991) reported that realistic root growth inhibition of *Lotus* could be obtained from a low ionic strength solution and a high Al concentration at similar levels to those in acid soils. Till now, comprehensive studies on the identification of the primary inhibitory factor for plant production are very limited. Specifically, clarification of plant nutritional characteristics for better plant production using two plant species, each composed of a wide variety of cultivars has only been reported by Akhter et al. (2009a).

Sorghum is the fifth most important cereal crop of the world (Sere and Estrada 1987). Although sorghum has been previously demonstrated to exhibit higher Al tolerance than wheat (Sasaki et al. 2006, Caniato et al. 2007), but recent studies on comparison of Al tolerance among rice, maize, wheat and sorghum indicated that both wheat and sorghum are very senstive to Al and Al tolerance of these two crop species are similar (Famoso et al. 2010). In South America, it is grown mainly on acid soils (4.6 million ha), and its production on these soils is limited by low content of cations, the toxicity of exchangeable Al and/or soluble Al and deficient levels of available P, Ca, Mg and micronutrients, and toxic levels of Al and Mn (Sere and Estrada 1987). A method to identify the primary soil factor from complicated tropical acid soil factors to evaluate acid soil adaptation is urgently needed for breeding for acid soil tolerance in crops. In this chapter, we aimed to clarify individual and combined tolerance to Al and low nutrient supply using a solution culture that simulates the nutrient status and concentration of soluble Al in tropical acid soils. Although sorghum is the primarily subject to be discussed in this chapter, maize, relatively superior in Al tolerance, is compared comprehensively as a reference crop with different cultivars.

ALUMINUM TOLERANCE SCREENING OF SORGHUM

Toxic symptoms of crop plant species to high Al are widely studied. Plant roots injured by high Al become stubby and thick, dark colored, brittle, poorly branched and suberized with a reduced root length and volume (Nguyen et al. 2001). Al can inhibit cytoskeletal dynamics and interacts with both microtubules and actin filaments (Sivaguru et al. 1999, Kochian et al. 2005) and this growth inhibition of root further cause reduced plant vigor and yield (Rengel, 1992, Kochian et al. 2005). Shoot growth is also inhibited due to limiting supply of water and nutrients. Al toxicity causes Ca deficiency or reduced Ca transport within the plant by curling or rolling of young leaves, inhibit growth of lateral branches or a collapse of growing points or petioles. Young seedlings are affected more than older plants (Thaworuwong and van Diest, 1974). Visible symptoms of Al toxicity include inhibition of root growth (Delhaize and Ryan, 1995), swelling of the root tip and/or sloughing off the epidermis, plasma membrane depolarization, alteration of Ca^{2+} fluxes at the root-tip, stimulation of callose deposition (Schreiner et al. 1994, Zhang et al. 1994) and induction of rigor in the actin cytoskeleton (Grabski and Schindler, 1995). Since, uptake of Al into root has been reported to be very rapid usually within tens of minutes and the search for the primary sites for Al toxicity and tolerance of Al has so far been elusive, the use of short-term screening techniques for differential Al tolerance, root elongation measurement has been the most popular and has been suggested to be used as a common method to quantify Al tolerance (Horst, 1995) in spite of it's inherent complexity (Rengel, 1996). Therefore, this trait has been widely used to quantify the level of Al-tolerance of plant species although concentration of Al used in the solution during the study of Al tolerance screening was not the same among researchers. For example, Wagatsuma et al (2005) screened 18 crop plant species, cultivars and lines using same Al concentration (20 µM $AlCl_3$ in 0.2mM $CaCl_2$). On the other hand, Ishikawa and Wagatsuma (1998) studied Al tolerance using 5 µM $AlCl_3$ for barley, maize, pea and rice cultivars. Ma et al. (2005) screened Al tolerance of rice mutants with 10 or 50 µM $AlCl_3$ solutions. Therefore, suitable concentration of Al for Al tolerance screening study of sorghum was under question. To resolve this problem, Akhter et al. (2009b) evaluated and justified Al tolerance screening techniques. They compared Al tolerance of several cultivars of sorghum and maize treating with two concentrations of Al solution (2.5 and 20 µM $AlCl_3$). Based on average values of Al tolerance, sorghum stands as Al-sensitive and maize as Al-tolerant crop. For sorghum, significant positive correlation was observed between the Al-tolerance in low and high Al (R^2 = 0.561**) (Figure 1A). Similar significant positive correlation was also observed for Al-tolerant maize (R^2 = 0.914**) (Figure 1B). Finally it was suggested that suitable concentration of Al to screen Al tolerance depends upon crop species and for Al sensitive crop plant species like sorghum, lower concentration of Al is better for screening.

STRESS FACTORS UNDER TROPICAL ACID SOIL CONDITIONS

Akhter et al. (2009a) defined three tolerances in tropical acid soil conditions viz. Al, low-nutrient and combined (Al with low-nutrient) stress. They defined Al tolerance in short-term (24h Al treatment) condition as the ratio of root elongation in Al to that in control. On the other hand, four stress/tolerance definitions were described for long-term (for one month) based on dry weight. Al tolerance in full nutrient condition was the ratio of dry weight in Al under full nutrient (FN) supply condition. Al tolerance in low nutrient condition was the ratio of dry weight in Al under low nutrient (LN) supply condition. Low nutrient tolerance was defined as the ratio of dry weight in low nutrient condition compared with full nutrient conditions. As predominant stresses in tropical acid soil are simultaneous effect of Al and low nutrient stress i.e. combined stress/tolerance. This combined tolerance was defined as the ratio of dry weight in Al and low-nutrient stress conditions with full nutrient conditions. These definitions are shown below in a simpler way –

$$\text{Al tolerance in FN (\%)} = \frac{\text{Dry weight in FN + Al}}{\text{Dry weight in FN}} \times 100$$

$$\text{Al tolerance in LN (\%)} = \frac{\text{Dry weight in LN + Al}}{\text{Dry weight in LN}} \times 100$$

$$\text{Low nutrient tolerance (\%)} = \frac{\text{Dry weight in LN}}{\text{Dry weight in FN}} \times 100$$

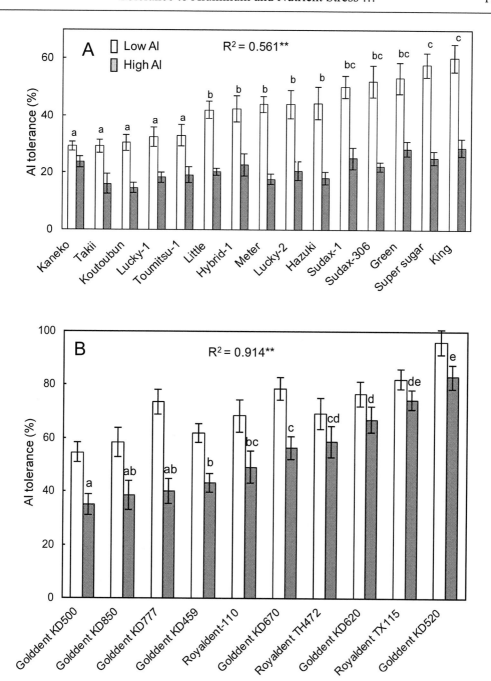

Figure 1. Differences in short-term aluminum (Al) tolerance among cultivars of sorghum (A) and maize (B). Low Al, 2.5 µM AlCl$_3$ in 0.2 mM CaCl$_2$ for 24 h (pH 5.0); high Al, 20 µM AlCl$_3$ in 0.2 mM CaCl$_2$ for 24 h (pH 4.9). Al tolerance is expressed as the net root elongation of the longest root in Al treatment/net root elongation of the control. Data are mean ±SE (n ≥ 10). Average values with the same letter(s) are not significantly different at the 5% significance level (Fisher's least significant difference). * $P < 0.05$; ** $P < 0.01$. The R^2 value is the determination coefficient between Al tolerance in low Al and that in high Al.

Table 1. Elemental composition (mM or μM) and pH of the long-term solution culture medium

Salt	Element	Unit	Treatments			
			Full nutrients (FN)		Low nutrients (LN)	
			FN	FN+Al	LN	LN+Al
NH_4NO_3	NH_4-N, NO_3-N		2.86	2.86	0.57	0.57
$NaNO_3$	NO_3-N		1.43	1.43	0.29	0.29
$NaH_2PO_4 \cdot 2H_2O$	P	mM	0.26	0.005	0.05	0.005
K_2SO_4	K		1.53	1.53	0.31	0.31
$CaCl_2 \cdot 2H_2O$	Ca		2.0	2.0	0.40	0.40
$MgSO_4 \cdot 7H_2O$	Mg		1.65	1.65	0.33	0.33
$FeSO_4 \cdot 7H_2O$	Fe		35.8	35.8	7.16	7.16
$MnSO_4 \cdot 5H_2O$	Mn		5.46	5.46	1.09	1.09
$CuSO_4 \cdot 5H_2O$	Cu		0.16	0.16	0.03	0.03
$(NH_4)_6Mo_7O_{24} \cdot 4H_2O$	Mo	μM	0.05	0.05	0.01	0.01
H_3BO_3	B		37.0	37.0	7.4	7.4
$ZnCl_2$	Zn		3.06	3.06	0.61	0.61
$AlCl_3 \cdot 6H_2O$	Al		—	11.1 (42.6)	—	11.1 (42.6)
pH			5.2	4.5 (4.3)	5.2	4.5 (4.3)

Except for P and Al, all nutrients are added concentration. P and Al are measured concentrations. Al concentration and pH in high Al conditions were described in parenthesis.

$$\text{Combined tolerance (\%)} = \frac{\text{Dry weight in LN + Al}}{\text{Dry weight in FN}} \times 100$$

In order to evaluate genotypic differences in tolerance to above defined stress factors, Afrin et al. (2009a) simulated an experimental design in the glasshouse mimicking low fertile tropical acid soils. Nutrient status and stress conditions in the culture solutions used were shown in Table 1.

SHORT-TERM ALUMINUM TOLERANCE VS. LONG-TERM ALUMINUM TOLERANCE

For sorghum, a wide range of variation in Al tolerance was observed among the cultivars which were similar to other Al tolerant crops such as maize. For numerous crop species including sorghum, short-term (24 h) screening technique for Al tolerance was suggested to be useful for estimating Al tolerance in long-term culturing with nutrients. The relationship between short-term and long-term Al tolerance was reported as low but significant (R^2 = 0.267*) for the whole plant under low-nutrient and low-Al conditions (Figure 2A). For maize, the relationship between short-term and long-term Al tolerance was high and significant ($R^2 = 0.462*$) for the whole plant under low-nutrient and high-Al conditions (Figure 2B). On the other hand, no correlations were found between short-term Al tolerance and the combined tolerance: $R^2 = 0.106$ and 0.002 for high-Al and low-Al conditions, respectively, for sorghum whereas the correlations between short-term Al tolerance and the combined tolerance were $R^2 = 0.172$ and 0.035, respectively (Akhter et al. 2009a). Although investigations based on similar short-term screening techniques have been reported (Ishikawa et al. 2001, Khan et al. 2009a, 2009b, Kobayashi et al. 2004, Ma et al. 2002, Wagatsuma et al. 2005, You et al. 2005), results presented in this chapter suggests that a short-term screening technique may not be particularly useful for estimating cultivar adaptation to the combination of stress factors found in tropical acid soils.

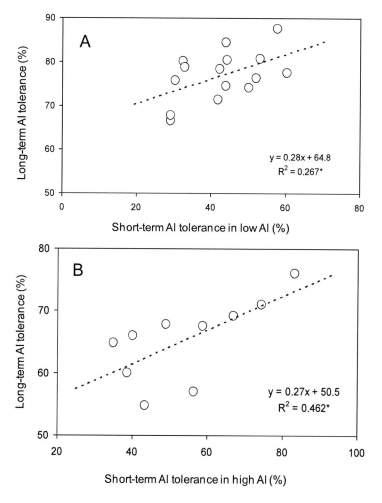

Figure 2. Relationship between short-term Al tolerance and long-term Al tolerance for sorghum (A) and maize (B). The Al tolerance for sorghum is under low Al conditions and that for maize under high Al conditions. * $P < 0.05$; ** $P < 0.01$.

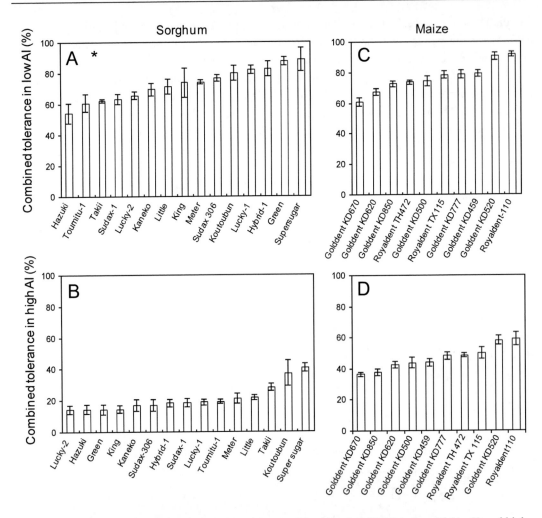

Figure 3. Combined stress tolerance (%) of sorghum (A, B) and maize (C, D) in low Al (A, C) and high Al (B, D) conditions. Tolerances were calculated based on the dry weight after culturing in different treatment solutions as shown in Table 1 for 29 d with a daily pH adjustment. The combined tolerance was defined as relative dry weight in LN+Al to that in FN. All tolerance values were calculated for the whole plant; * the combined tolerance of sorghum under low Al conditions for the shoot part only.

ALUMINUM TOLERANCE UNDER LOW-NUTRIENT CONDITIONS

Akhter et al. (2009a) cultured sorghum in low and high Al conditions with maize as a reference crop. They found that in low Al conditions the combined stress tolerance for sorghum shoots was in the range of 54.0-88.8% and mean value was 72.8 and under high Al conditions (sorghum whole plant) tolerance ranged from 14.2 to 40.7% with a mean value of 21% (Figure 3A, B). On the other hand, combined stress tolerance of the reference crop, maize (whole plant), under low Al conditions ranged from 60.8 to 91.7% with a mean value of 76.8% and that under high Al conditions ranged from 36.5 to 59.1% with a mean value of 46.9% (Figure 3C, D).

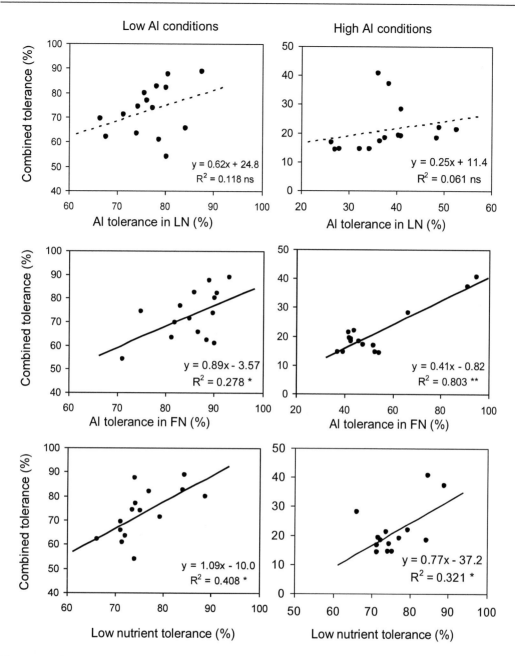

Figure 4. Relationship for Al tolerance in LN, Al tolerance in FN and low-nutrient tolerance with combined stress tolerance for sorghum in low Al and high Al conditions. Tolerances were calculated based on the dry weight after culturing in different treatment solutions as shown in Table 1 for 29 d with a daily pH adjustment. Combined tolerance has been defined in Figure 2. Al tolerance in LN, relative dry weight in LN+Al to that in LN; Al tolerance in FN, relative dry weight in FN+Al to that in FN; LN tolerance, relative dry weight in LN to that in FN. Dotted lines indicate non significant relations. ns, not significant; * $P < 0.05$; ** $P < 0.01$.

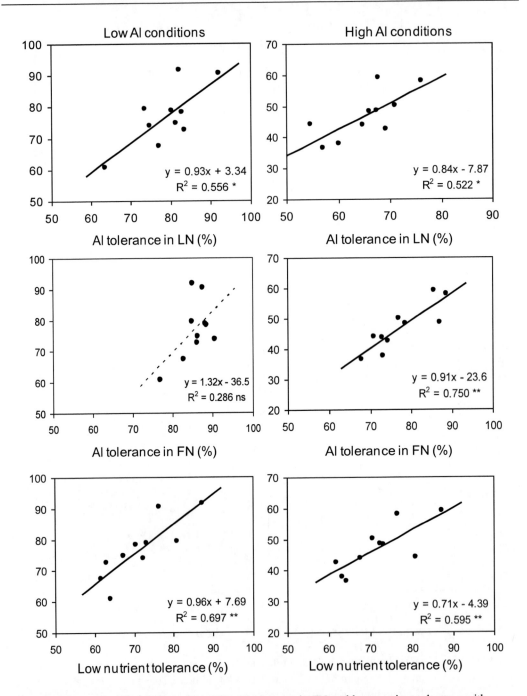

Figure 5. Relationship for Al tolerance in LN, Al tolerance in FN and low-nutrient tolerance with combined stress tolerance for maize in low Al and high Al conditions. Treatments conditions, materials and definitions are same in Figure 3. All tolerance values were calculated for the whole plant. Dotted line indicates non significant relation. ns, not significant; * $P < 0.05$; ** $P < 0.01$.

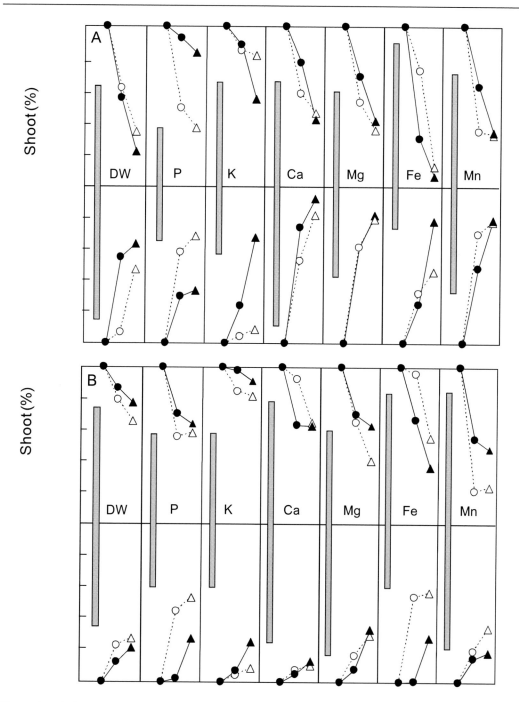

Figure 6. Relative values on average values of dry weight and mineral concentration for 15 cultivars of sorghum (A) and 10 cultivars of maize (B) grown in long-term culturing. ▨ dry weight or mineral concentration in LN (% of FN), ●— LN under low Al conditions (% of LN), ▲— LN under high Al conditions (% of LN), ⋯○⋯ FN under low Al conditions (% of FN), ⋯△⋯ FN under high Al conditions (% of FN).

Table 2. Multiple regression equation (M), standardized partial regression equation (S) and probability (P) on the relationship between combined tolerance, Al tolerance and low nutrient tolerance

Plant species	Al treatment	Equation and probability
Sorghum	Low Al [a]	M:CT = − 56 + 0.661 AT + 0.857 LNT S:CT = 0.39 AT + 0.46 LNT $P = 0.023$
	High Al	M:CT = − 4.2 + 0.301 AT + 0.121 LNT S:CT = 0.67 AT + 0.02 LNT $P < 0.0001$
Maize	Low Al	M:CT = − 56 + 0.706 AT + 0.794 LNT S:CT = 0.57 AT + 0.69 LNT $P = 0.0019$
	High Al	M:CT = − 47 + 0.739 AT + 0.643 LNT S:CT = 0.64 AT + 0.69 LNT $P = 0.0006$

[a] Calculation on sorghum in full nutrient with low Al conditions only was on shoot part. All others are on whole plant in low nutrient.
CT, Combined tolerance; AT, Al tolerance; LNT, Low nutrient tolerance.

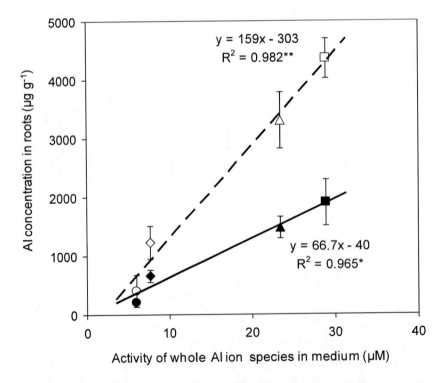

Figure 7. Relationship between activity of total Al ion species in medium and average Al concentration in roots for all sorghum cultivars (open symbols) and all maize cultivars (closed symbols). ○, ●, full nutrient + low Al; ◇, ◆, full nutrient + high Al; △, ▲, low nutrient + low Al; □, ■, low nutrient + high Al. Values are mean ± SE (n = 3).

In all experimental conditions, the combined stress tolerance was significantly correlated with Al tolerance and low nutrient tolerance for both sorghum and maize cultivars (Figures 3, 4, 5 and Table 2). To evaluate the relative importance of these two factors on the combined stress tolerance, Akhter et al. (2009a) conducted a multiple regression analysis and further calculated a standardized partial regression equation (to treat both factors equally) among the combined stress tolerance, Al tolerance and low nutrient stress tolerance (Table 2). Although these results did not show any correlation of combined stress tolerance for sorghum whole plants in the low Al conditions, the shoot (i.e. the harvested plant parts) growth showed a relationship with the other tolerances that constituted almost three-quarters of the whole plant. A greater contribution of low nutrient stress tolerance than Al tolerance under most conditions was evident (0.46>0.39 for sorghum under low Al conditions, 0.69>0.57 for maize under low Al conditions and 0.69>0.64 for maize under high Al conditions). For sorghum, Al tolerance is important only in high Al conditions, however, for maize low nutrient stress tolerance is important irrespective of Al conditions. Although similar results have already been reported (Okada and Fischer 2001, Wenzl et al. 2003), findings of Akhter et al. (2009b) can be considered definitive evidence because it is based on a more comprehensive experiment.

Presence of Al in the nutrient solutions decreased the dry weight production and also decreased the concentration of all of the nutrients measured (P, K, Ca, Mg, Fe and Mn) in sorghum more than in maize (Figure 6). A less inhibitory effect of Al on the dry weight production in maize indicates greater Al tolerance compared with sorghum. The activity of all Al ion species ($Al[OH]_3^0 + AlSO_4^+ + Al[OH]_2^+ + AlOH_2^+ + Al^{3+}$) in the low nutrient medium was higher than that in the full nutrient medium as determined by the calculation of Wada and Seki (1994). The equation for the regression line of the mean Al concentration in the roots of all cultivars ($\mu g\ g^{-1}$ [y]) and the activity of the Al ions in the medium (x) for sorghum under four different medium conditions, i.e., full nutrient + low Al, full nutrient + high Al, low nutrient + low Al and low nutrient + high Al, was y = 159x − 303 ($R^2 = 0.982**$) (Al concentrations ranged between 402 and 4359 $\mu g\ g^{-1}$). The equation for the regression line between the above two factors for maize was y = 66.7x − 40 ($R^2 = 0.965*$) (Al concentrations ranged between 202 and 1905 $\mu g\ g^{-1}$) (Figure 7). Comparing each slope of the regression line for sorghum and maize, i.e., 159 and 66.7, the absorbability of the Al ions in sorghum roots was estimated to be more than twice to that of maize roots. Similar higher Al^{3+} activity was also observed with the decrease of ionic activity of other cations in the solution (Kinraide and Parker, 1987). Comparative ameliorative effect of Ca and other cations have been revealed in the following order: $Ca^{2+} \approx Mg^{2+} \approx Sr^{2+} > Na^+ \approx K^+$ (Kinraide and Parker, 1987; Kinraide et al., 1992, 1994; Ryan et al., 1994, 1997). Although comparative effectivness among monovalent and divalent cations are not so clear, it is obvious that divalent cations are more effective amelionats for Al than monovalent cations.

RELATIONSHIP BETWEEN ALUMINUM TOLERANCE AND MINERAL ELEMENT STATUS

For sorghum, the Al in the roots was negatively correlated with the Al tolerance in the low nutrient treatment under high Al conditions ($R^2 = 0.278*$) (Figure 8A). This indicates that the lower level of Al tolerance for sorghum is the result from the greater absorbability of Al ions by roots. This negative correlation between Al tolerance and Al concentrations agrees with previous reports (Khan et al. 2009a, 2009b; Ofei-Manu et al. 2001, Pineros et al. 2005, Wagatsuma et al. 1995). On the other hand, shoot K concentration of sorghum was positively correlated with Al tolerance in the low nutrient treatment (LN) under high Al conditions (HA) ($R^2 = 0.284^*$) (Figure 8B). In sorghum, higher K in shoot in (LN + HA) is associated with higher combined stress tolerance of shoot ($R^2 = 0.491^{**}$) (Figure 9A); in maize, higher shoot Ca in LN + HA is connected with higher combined tolerance of shoot ($R^2 = 0.477^*$) (Figure 9B).

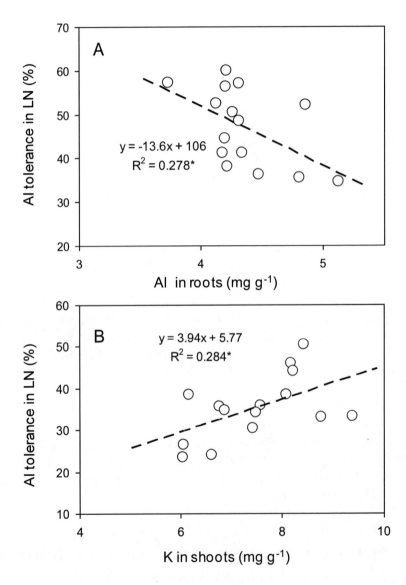

Figure 8. Relationship between Al tolerance in LN and Al concentration in roots (A) or between Al tolerance and K concentration in shoots (B) for sorghum in long-term culturing. Al tolerance in LN was calculated as the ratio of growth in LN under high Al conditions to that in LN. * $P < 0.05$.

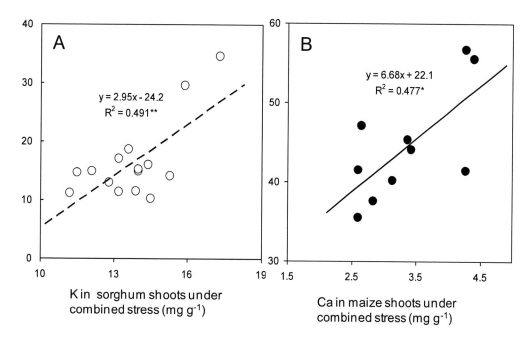

Figure 9. Relationship between combined tolerance and K concentration for sorghum shoots (A) and between the Ca concentrations for maize shoots (B) in LN under high Al conditions. * $P < 0.05$; ** $P < 0.01$.

Although the soluble P concentration used was rather low (5 µmol L^{-1}) in the results presented in this chapter, no positive correlations were observed between shoot P and any type of tolerance at shoot level (R^2 = 0.116, 0.017, 0.166, 0.191 and 0.256 for Al tolerance [low Al], Al tolerance [high Al], low nutrient stress tolerance, combined stress tolerance [low Al] and combined tolerance [high Al] in sorghum respectively; and R^2 = 0.095, 0.334, 0.061, 0.037 and 0.396 for Al tolerance [low Al], Al tolerance [high Al], low nutrient tolerance, combined tolerance [low Al] and combined tolerance [high Al] in maize, respectively) (data not shown). The soluble concentration of P in the medium is, therefore, not considered to be a determining factor for sorghum growth in the present medium under simulated tropical acid soil condition.

CONCLUSION

Studies on the clarification of primary growth limiting factors in tropical acid soils are limited. Research on Al tolerance would be useful for plant breeders to make Al tolerant crops. However, tropical acid soil is different from general acid soil. Tropical acid soils consists of high Al and low nutrient stresses and research on this soil should be carried out considering these two factors simultaneously. This chapter discusses a new aspect for the plant breeders to identify genotypes with tolerance to combined stress factors of Al toxicity and low nutrient supply. Also, this will be a new feature for the farmer to choose a crop species or cultivar for his farm and make crop cultivation profitable.

This chapter propose that the plant nutritional characteristics linked to low nutrient stress tolerance demonstrated in the present investigation should be evaluated as an important strategy for plant production in tropical acid soils in both Al tolerant and Al sensitive plant species under low Al conditions. For sensitive crop species (sorghum) in infertile acid soil, Al toxicity would be a major factor to regulate production but for Al-tolerant crop species (maize), low nutrient tolerance would be more important factor than Al tolerance. Here, for the first time, we clarified that primary growth limiting factors depend on the crop species, concentration of Al and ionic strength of the solution. Further research is needed on the above point using many cultivars with wide variation of Al tolerance for more widely used crops.

REFERENCES

Akhter, A., Khan, M.S.H., Egashira, H., Tawaraya, K., Rao, I.M., Wenzl, P., Ishikawa, S., and Wagatsuma, T. (2009a). Greater contribution of low-nutrient tolerance to the growths of sorghum and maize under combined stress conditions with high-aluminum and low-nutrients in solution culture simulating the nutrient status of tropical acid soils. *Soil Sci. Plant Nutr.*, 55, 394-406.

Akhter, A., Wagatsuma, T., Khan, M.S.H., and Tawaraya, K. (2009b). Comparative studies on aluminum tolerance screening techniques for sorghum, soybean and maize in simple solution culture. *Am. J. Plant Physiol.*, 4, 1-8.

Blamey, F.P.C., Edmeades, D.C.; Asher, C.J., Edwards, D.G., and Wheeler, D.M. (1991): Evaluation of solution culture techniques for studying aluminum toxicity in plants. In: R.J. Wright et al. (Eds) *Plant–Soil Interactions at Low pH,* pp. 905–912. Kluwer Academic Publishers, the Netherlands.

Blamey, F.P.C., Edwards, D.G.; Asher, C.J. (1983). Effect of aluminum, OH:Al and P:Al ratios, and ionic strength on soybean root elongation in solution culture. *Soil Sci.*, 136, 197–207.

Caniato, F.F., Guimaraes, C.T., Schaffert, R.E., Alves, V.M., Kochian, L.V., Borem, A., Klein, P.E., and Magalhaes, J.V. (2007). Genetic diversity for aluminum tolerance in sorghum. *Theor. Appl. Genet.*, 114, 863-876.

Delhaize, E., and Ryan, P.R. (1995). Aluminum toxicity and tolerance in plants. *Plant Physiol.*, 107, 315-321.

Famoso, A.N., Clark, R.T., Shaff, J.E., Craft, E., McCough, S.R., and Kochian, L.V. (2010). Development of a novel aluminum tolerance phenotyping platform used for comparisons of cereal aluminum tolerance and investigations into rice aluminum tolerance mechanisms. *Plant Physiol.*, 153, 1678-1691.

Foy, C.D., and Brown, J.C. (1964) Toxic factors in acid soils. II. Differential aluminum tolerance of plant sciences. *Soil Sci. Soc. Am. Proc.*, 28, 27–32.

Grabski, S., and Schindler, M. (1995). Aluminum induces rigor within the actin network of soybean cells. *Plant Physiol.*, 108, 897-901.

Hãussler, K., Rao, I.M., Schultzekraft, R., and the late Marschner, H. (2006), Shoot and root growth of two tropical grasses, *Bracharia ruziziensis* and *B. dictyoneura,* as influenced by aluminum toxicity and phosphorus deficiency in a sandy loam Oxisol of the eastern plains of Colombia. *Tropical Grassland*, 40, 213–221.

Horst, W.J. (1995). The role of the apoplast in Al toxicity and resistance of higher plants. *Z. Pflanzenernahr. Bodenk.*, 158, 419-428.

Ishikawa, S., and Wagatsuma, T. (1998). Plasma membrane permeability of root-tip cells following temporary exposure to Al ions is a rapid measure of Al tolerance among plant species. *Plant Cell Physiol.*, 39, 516-525.

Ishikawa, S., Wagatsuma, T., Takano, T., Tawaraya, K., and Oomata, K. (2001). The plasma membrane intactness of root-tip cells is a primary factor for Al-tolerance in cultivars of five species. *Soil Sci. Plant Nutr.*, 47, 489–501.

Khan, M.S.H., Tawaraya, K., Sekimoto, H., Koyama, H., Kobayashi, Y., Murayama, T., Chuba, M., Kambayashi, M., Shiono, Y., Uemura, M., Ishikawa, S., and Wagatsuma, T. (2009a). Relative abundance of Δ^5-sterols in plasma membrane lipids of root- tip cells correlates with aluminum tolerance of rice. *Physiol. Plant.*, 135, 73-83.

Khan, M.S.H., Wagatsuma, T., Akhter, A., and Tawaraya, K. (2009b). Sterol biosynthesis inhibition by paclobutrazol induces greater aluminum (Al) sensitivity in Al-tolerant rice. *Am. J. Plant Physiol.*, 4, 89-99.

Kinraide, T.B., and Parker, D.R. (1987). Cation amelioration of aluminum toxicity in wheat. *Plant Physiol.*, 83, 546–551.

Kinraide, T.B., Ryan, P.R., and Kochian, L.V. (1992). Interactive effects of Al^{3+}, H^+, and other cations on root elongation considered in terms of cell-surface electrical potential. *Plant Physiol.*, 99, 1461–1468.

Kinraide, T.B., Ryan, P.R., and Kochian, L.V. (1994). Al^{3+}-Ca^{2+} interactions in aluminum rhizotoxicity. II. Evaluating the Ca^{2+}-displacement hypothesis. *Planta*, 192, 104–109.

Kobayashi, Y., Yamamoto, Y., and Matsumoto, H. (2004). Studies on the mechanism of aluminum tolerance in pea (*Pisum sativum* L.) using aluminum-tolerant cultivar 'Alaska' and aluminum-sensitive cultivar 'Hyogo'. *Soil Sci. Plant Nutr.*, 50, 197–204.

Kochian, L.V., Pineros, M.A., and Hoekenga, O.A. (2005). The physiology, genetics and molecular biology of plant aluminum resistance and toxicity. *Plant Soil*, 274, 175-195.

Ma, J.F, Nagao, S., Huang, C.F, and Nishimura, M. (2005). Isolation and characterization of rice mutant hyper sensitive to Al. *Plant Cell Physiol.*, 46, 1054-1061.

Ma, J.F., Shen, R., Zhao, Z., Wissuwa, M., Takeuchi, Y., Ebitani, T., and Yano, M. (2002). Response of rice to Al stress and identification of quantitative trait loci for Al tolerance. *Plant Cell Physiol.*, 43, 652–659.

Mariano, E.D., and Keltjens, W. (2005). Long-term effects of aluminum exposure on nutrient uptake by maize genotypes differing in aluminum resistance. *J. Plant Nutr.*, 28, 323–333.

Nguyen, V.T., Burow, M.D., Nguyen, H.T., Le, B.T., Le, T.D., and Paterson, A.H. (2001). Molecular mapping of genes conferring aluminum tolerance in rice (*Oryza sativa* L.) *Theor. Applied Genet.*, 102, 1002–1010.

Ofei-Manu, P., Wagatsuma, T., Ishikawa, S., and Tawaraya, K. (2001). The plasma membrane strength of the root-tip cells and root phenolic compounds are correlated with Al tolerance in several common woody plants. *Soil Sci. Plant Nutr.*, 47, 359–375.

Okada, K., and Fischer, A.J. (2001). Adaptation mechanism of upland rice genotypes of highly weathered acid soils of South American Savannas. In: N. Ae et al. (eds) Plant Nutrient Acquisition: New perspectives, pp. 185–200, NIAES Series 4, Springer, Tokyo.

Pavan, M.A., Bingham, F.T., and Pratt, P.F. (1982). Toxicity of aluminum to coffee in Ultisols and Oxisols amended with $CaCO_3$, $MgCO_3$ and $CaSO_4 \cdot 2H_2O$. *Soil Sci. Soc. Am. J.*, 46, 1201-1207.

Piñeros, M.A., Shaff, J.E., Manslank, H.S., Alves, V.M.C., and Kochian, L.V. (2005). Aluminum resistance in maize cannot be solely explained by root organic acid exudation. A comparative physiological study. *Plant Physiol.*, 137, 231-241.

Pintro, J., Inoue, T.T., and Tescaro, M.D. (1999). Influence of the ionic strength of nutrient solutions and tropical acid soil solutions on aluminum activity. *J. Plant Nutr.*, 22, 1211-1221.

Pintro, J.C., and Taylor, G.J. (2004). Effects of aluminum toxicity on wheat plants cultivated under conditions of varying ionic strength. *J. Plant Nutr.*, 27, 907–919.

Rao, I.M. (2001). Adapting tropical forages to low-fertility soils. *Proceedings of the XIX International Grassland Congress*, pp. 247–254. São Pedro-SP.

Rao, I.M., Zeeigler, R.S., Vera, R., and Sarkarung, S. (1993). Selection and breeding for acid soil tolerance in crops. Upland rice and tropical forages as case studies. *BioScience*, 43, 454–465.

Rengel, Z. (1992). Disturbance of cell Ca^{2+} homeostasis as primary trigger of Al toxicity syndrome. *Plant Cell Environ.*, 15, 931-938.

Rengel, Z. (1996). Uptake of aluminum in plant cells. *New Phytol.*, 134, 389-406.

Ryan, P.R., Kinraide, T.B., and Kochian, L.V. (1994). Al^{3+}-Ca^{2+} interactions in aluminum rhizotoxicity. I. Inhibition of root growth is not caused by reduction of calcium uptake. *Planta*, 192, 98–103.

Ryan, P.R., Reid, R.J., and Smith, F.A. (1997). Direct evaluation of the Ca^{2+}-displacement hypothesis for Al toxicity. *Plant Physiol.*, 113, 1351–1357.

Sanchez, P.A. and Salinas J.G. (1981). Low-input technology for managing oxisols and ultisols in tropical America. *Adv. Agron.*, 34, 279-406.

Sasaki, T., Ryan, P.R., Delhaize, E., Hebb, D.M., Ogihara, Y., Kawarura, K., Noda, K., Kojima, T., Toyoda, A., Matsumoto, H., and Yamamoto, Y. (2006). Sequence upstream of the wheat (*Triticm aestivum* L.) ALMT1 gene and its relationship to aluminum resistance. *Plant Cell Physiol.*, 47, 1343-1354.

Schreiner, K.A., Hoddinott, J., and Taylor, G.J. (1994). Aluminum-induced deposition of (1,3,)-Betaglucans (callose) in *Triticum aestivum* L. *Plant Soil*, 162, 273-280.

Sere, C., and Estrada, R.D. (1987). Potential role of grain sorghum in the agricultural systems of regions with acid soils in tropical Latin America. In: L.M. Gourley and J.G. Salinas (Eds) *Sorghum for Acid Soils*. pp. 145–169. International Center for Tropical Agriculture (CIAT), Cali, Colombia.

Sivaguru, M., Baluska, F., Volkmann, D., Felle, H.H., and Horst, W.J. (1999). Impacts of aluminum on the cytoskeleton of the maize root apex. Short-term effects on the distal part of the transition zone. *Plant Physiol.*, 119, 1073-1082.

Sumner, M.E., and Noble, A.D. (2003). Soil acidification: the world story. In: Z. Rengel (Ed.) *Handbook of soil acidity*. pp. 1-28. Marcel Dekker, Inc., New York.

Thaworuwong, N., and van Diest. (1974). Influence of high acidity and aluminum on the growth of lowland rice. *Plant Soil*, 41, 141-159.

Wada, S.-I., and Seki, H. (1994). A compact computer code for ion speciation in aqueous solutions based on a robust algorithm. *Soil Sci. Plant Nutr.*, 40, 165–172.

Wagatsuma, T., Ishikawa, S., Obata, H., Tawaraya, K., and Katohda, S. (1995). Plasma membrane of younger and outer cells is the primary specific site for aluminum toxicity in roots. *Plant Soil*, 171, 105–112.

Wagatsuma, T., Khan, M.S.H., Rao, I.M., Wenzl, P., Tawaraya, K., Yamamoto, T., Kawamura, T., Hosogoe, K., and Ishikawa, S. (2005). Methylene blue stainability of root-tip protoplasts as an indicator of aluminum tolerance in a wide range of plant species, cultivars and lines. *Soil Sci. Plant Nutr.*, 51, 991-998.

Wenzl, P., Mancilla, L.I., Mayer, J.E., Albert, R., and Rao, I.M. (2003). Simulating infertile acid soils with nutrient solutions: the effect on *Brachiaria* species. *Soil. Sci. Soc. Am. J.*, 67, 1457–1469.

Wenzl, P., Patiño, G.M., Chaves, A.L., Mayer, J.E., and Rao, I.M. (2001). The high level of aluminum resistance in signalgrass is not associated with known mechanisms of external aluminum detoxification in root apices. *Plant Physiol.*, 125, 1473–1484.

Wheeler, D.M., and Edmeads, D.C. (1995). Effect of ionic strength on wheat yield in the presence and absence of aluminum. In: R.A. Date et al. (Eds) *Plant–soil interactions at low pH*, pp. 623–626. Kluwer Academic Publishers, Dordrecht.

Wright, R.J., Baligar, V.C., and Ahlrichs, J.L. (1989). The influence of extractable and soil solution aluminum on root-growth of wheat seedlings. *Soil Sci.*, 148, 293–302.

You, J.F., He, Y.F., Yang, J.L., and Zheng, S.J. (2005). A comparison of aluminum resistance among *Polygonum* species originating on strongly acidic and neutral soils. *Plant Soil*, 276, 143–151.

Zhang, G.C., Hoddinott, J., and Taylor, G.J. (1994). Characterization of 1,3-beta-d-glucan (callose) synthesis in roots of *Triticum aestivum* in response to aluminum toxicity. *J. Plant Physiol.*, 144, 229-234.

Reviewed by:
Hiroyuki Koyama PhD. Professor, Plant Cell Technology; Applied Biological Sciences; Gifu University. 1-1, Yanagido, 501-1193, Gifu, Japan. Email: koyama@gifu-u.ac.jp

Chapter 7

BIOLOGICAL ACTIVITIES OF SORGHUM EXTRACT AND ITS EFFECT ON ANTIBIOTIC RESISTANCE

Silvia Mošovská, Lucia Birošová, Ľubomír Valík
Department of Nutrition and Food Assessment, Faculty of Chemical and Food Technology, Slovak University of Technology, Radlinského 9, 812 37 Bratislava, Slovak Republic

ABSTRACT

Sorghum is an ancient crop belonging to the gluten-free cereals that are known as pseudocereals. It is an important food in semi-arid tropics of the world including Africa or Asia because sorghum as other pseudocereals is extensively drought tolerant and well-adapted to weather extremes. It is quite widely used. It is applied as a forage crop or is a raw material for bio ethanol and other industrial products. At present, the interest in sorghum is growing considerably as a potential crop in food industries. Since sorghum does not contain gluten it is suitable for people with celiac diseases. Seeds of sorghum are also a good source of various phytochemicals such as phenolic compounds including phenolic acids, flavonoids and condensed tannins which have potential biological activities. The objective of this study was to investigate antioxidant and antimicrobial activity of sorghum extract. The ability of extract to quench free radicals was measured by spectrophotometric ABTS (IC_{50} 0,811 mg/ml), DPPH (IC_{50} 1,645 mg/ml) and FRAP (IC_{50} 0,569 mg/ml) methods. Extract inhibited growth of fungi *Aspergillus flavus* and *Alternaria alternata* and showed also antibacterial effect on gram-positive (*Bacillus subtilis, Staphylococcus aureus, Staphylococcus epidermidis*) and gram-negative (*Salmonella typhimurium, Pseudomonas aeruginosa, Escherichia coli, Enterobacter sakazakii*) bacteria. In the last years, the frequency of antimicrobial resistant infection has increased and thus antibacterial resistance has become serious problem. The various experimental studies suggested that antimutagens could have important function in war against microbial resistance. Whereas our previous results indicated potential antimutagenic activity of tested extract, effect of sorghum on the development of antibiotic resistance was also studied.

Keywords: sorghum, polyphenolics, resistance.

ABBREVIATIONS

ABTS	2,2`-azinobis(3-ethylbenthiazoline-6-sulfonic acid;
DPPH	1,1-diphenyl-2-picrylhydrazyl;
FRAP assay	ferricreducing antioxidant power;
BHT	butylated hydroxytoluene;
TLC	thin layer chromatography

INTRODUCTION

Sorghum belongs to main grains in tropical and sub-tropical regions, especially in Africa, [1; 2] which are characterized by low rainfall, drought, high temperature and low soil fertility [3]. Since sorghum does not contain gluten it is suitable for diet of celiac disease [4]. The sorghum seed is a caryopsis which consists of pericarp, endosperm, and embryo. The endosperm is the major part of caryopsis and it is high in starch and proteins [1; 5]. There also are several sorghum varieties with different colour of testa which can be white, gray, red or brown in colour [5]. Sorghum has high protein content compared with corn, although nutritical quality of these proteins is a lower [6]. Nevertheless, sorghum is the important source of phytochemicals such as polyphenols, plant sterols and polycosanols which have positive effect on human health [7; 8]. Many epidemiological studies have suggested that increased consummation of whole grain product, fruit and vegetables due to reduction risk of chronic diseases. This effect is probably caused by natural antioxidants, including vitamin C, tocopherol, carotenoids and polyphenols, which protect the body against damage of free radicals [7]. The present polyphenolic compounds in sorghum occur in two major categories including flavonoids and phenolic acids. Phenolic acids include two classes, hydroxybenzoic and hydroxycinnamic acids, which are located in the pericarp, testa, aleurone layer, and endosperm. The major flavonoid group in sorghum varieties is condensed tannins and anthocyanins which content depend on the presence of a pigmented testa. In general, sorghum varieties which do not have a pigmented testa contain low levels of phenols and no tannins [8; 9; 10]. The majority of polyphenolics have displayed, except for antioxidant activity, antimicrobial and antimutagenic effect [11]. Since, food spoilage is one of the most important problem in the food industry [12] the applying of preservatives is still actual. In the last years, interest of scientists in various natural substances has grown for their potential biological, therapeutic and pharmacological properties [13]. Polyphenolic compounds are also interest for their potential antimutagenic effect. According to Pillai et al. [14] a number antimutagens are able to suppress the emergence of resistance. Birošová et al. [15] have found out that some phenolic acid like vanillic acid could decrease frequency of spontaneous and induced mutations leading to antibiotic resistance in *Salmonella typhimurium*.

The objective of this work was to study antioxidant activity by three independent methods (ABTS, DPPH and FRAP assay), inhibitory effect on some undesirable microorganisms in food and influence on the development of antibiotic resistance of sorghum extract.

MATERIAL AND METHODS

Bacterial Strains and Antibiotic

Bacterial strains of *Bacillus subtilis* (CCM 1999), *Staphylococcus aureus* (CCM 3953), *Staphyloccoccus epidermidis* (CCM 4418), *Salmonella enterica* subs. *enterica* serotype Typhimurium (CCM 4763), *Escherichia coli* (CCM 3954), *Pseudomonas aeruginosa* (CCM 3955), yeast *Saccharomyces cerevisiae* (CCM 8191), *Candida albicans* (CCM 8188) and fungi *Ryzopus oryzae* (CCM 8076), *Aspergillus flavus* (CCM 8363), *Fusarium culmorum* (CCM F-163), *Alternaria alternata* (CCM 8326) were obtained from Czech Collection of Microorganisms (Masaryk University, Brno, Czech Republic). *Candida maltosa* YP1 was obtained from institute of biochemistry, nutrition and health protection (Faculty of chemical and food technology, Slovak technical university, Bratislava, Slovakia).

Ciprofloxacin and gentamicin were purchased from Merck (Germany). Synthetic antioxidants BHT, ascorbic acid were bought from Sigma-Aldrich (Germany) and α-tocopherol was obtained from Merck (Germany).

Plant Material and Extraction

The sorghum seeds (*Sorghum bicolor* L.) variety Zsófia was obtained from Plant Production Station in Uhříněves (Czech Republic).

Sorghum extract was prepared according to Krygier et al. [16]. Briefly, 10 g defatted sorghum flour [17] was hydrolyzed with 100 ml 2 M NaOH for 4 h at 50 °C. After hydrolysis, mixture was acidified with 6 M HCl to pH 2 and supernatant was extracted with ethyl acetate at ratio 1:1 (v/v) 6 times. Ethyl acetate was evaporated in a rotary evaporator at temperature lower than 40 °C and the residue was dissolved in methanol (bioautography detection). For another analyses methanol was evaporated and then dissolved in DMSO for studying antioxidant, antimicrobial effects and determination of mutation frequencies leading to ciprofloxacin resistance.

ABTS Assay

The total antioxidant activity of sorghum extract on ABTS radical was established using the method of Arts et al. [18]. Briefly, at first the absorbance spectrum of ABTS radical cation was measured at the wavelenght from 730 to 1100 nm. Then, 50 µl sample was added to the cation and the absorbance spectrum was estimated after 10 minutes at the same wavelenghts as radical cation. The total antioxidant capacity of sorghum extract was expressed as IC 50 value in mg/ml and compared with synthetic antioxidants (BHT, ascorbic acid and α-tocopherol). Inhibition of free radical measured by ABTS test in percent of inhibition was calculated according to formula % AA = {[$A_{730ABTS}$ - ($A_{730sample}$ + $A_{1100ABTS}$ − $A_{1100sample}$)]/$A_{730ABTS}$}x100 where $A_{730ABTS}$ represents the absorbance of $ABTS^{\cdot+}$ at 730 nm, $A_{730sample}$ is the absorbance of sample at 730 nm, $A_{1100ABTS}$ expresses the absorbance of $ABTS^{\cdot+}$ at 1100 nm and $A_{1100sample}$ is the absorbance of sample at 1100 nm. Extract concentration providing 50 % inhibition was also calculated from the plot of inhibition against extract concentration. The results represent the mean of two experiments.

DPPH Assay

DPPH radical scavenging activity was measured according to Yen et al. [19]. DPPH radical was prepared by mixing 0,012 g DPPH and 100 ml 96% ethanol. For measurement of sample, 1 ml extract was pipeted into tubes, to which 3 ml ethanol and 1 ml DPPH radical was added. The absorbance of mixture was measured at the wavelenght of 517 nm after 10 minutes. Decrease of absorbance indicated the accurance of antioxidants free radicals quenching in sample. The antioxidant capacities of sorghum extract was also estimated as in previous occurrence. Inhibition of free radical measured by DPPH test in percent of inhibition was calculated according to formula % inhibition = [(A_{blank} - A_{sample})/ A_{blank}]x100 where A_{blank} is the absorbance of the control reaction and A_{sample} is the absorbance of the test compound. IC 50 value expressing extract concentration providing 50% inhibition was calculated from the plot of inhibition against extract concentration. The results represent the mean of two experiments.

FRAP Assay

FRAP assay was carried on method of Niemeyer and Metzler [20] with slight modification. For measurement, 60 µl sample in 180 µl water was added to 1800 µl FRAP reagent. The absorbance of mixture was measured at the wavelenght of 595 nm after 30 minutes. The antioxidant effect was also estimated as in ABTS and DPPH test. Reduction of $Fe(3+)$ to $Fe(2+)$ carried out by FRAP method in percent of inhibition was evaluated in following way % reduction = [(A_{sample} - A_{blank})/ A_{sample}] x100 where A_{blank} is the absorbance of the control reaction and A_{sample} is the absorbance of the test compound. Extract concentration providing 50% reduction was calculated from the plot of inhibition against extract concentration.

The Determination of Antimicrobial Activity

The antimicrobial activity of sorghum extract was determined by paper disk diffusion method [21]. Extracts in four tested concentration (6,25 – 50 mg/ml) were applied on disk placed on Muller-Hinton agar inoculated by tested microorganism. The plates were incubated for 24 h, after which the diameter of each inhibitory zone was evaluated and compared with reference antibiotic disc. Negative control was prepared using the same solvent in which extract was dissolved.

TLC Bioautography Analysis

The antimicrobial activity of individual component of extract was tested by TLC bioautography analysis according to Gu et al. [22] with some modifications on bacterial strains *B. subtilis* and *S. typhimurium*. The aliquot number of tested extract (3 μl) was deposited onto the TLC plates. The TLC plates were development in a presaturated solvent chamber with A – ethyl acetate:hexane (2:1), B – ethyl acetate:hexane (1:2) and C – toluene:ethyl acetate: formic acid (5:4:1). The developed TLC plates were monitored under UV light at 300nm and extract components were evaluated by retention factor (R_f). Retention factor was calculated according to formula $R_f = a/b$ where *a* represents distance traveled by the compound and *b* is distance traveled by the solvent front. Next, they were press down with sorbent page on Muller-Hinton agar plates inculated by tested microorganism. The biological active components diffused into agar plates for 30 minutes. Then, the TLC plates were removed and after incubation for 24 h, the antimicrobial effect of active metabolites displaying formation of inhibitory zone was evaluated.

Determination of Sorghum Effect on the Development of Ciprofloxacin Resistance

The influence of sorghum extract on the development of ciprofloxacin resistance was tested according to Birošová et al. [15]. Briefly, 0,1 ml of tested extract was divided into sterile tubes to which was added 0,5 ml of phosphate buffer (pH 7,4) and 0,1 ml of overnight culture *S. enterica* ser. Typhimurium (cell density of 10^9/ml). The mixture was treated for 30 min at 37 °C. After addition of 0,7 ml of Nutrient broth No.2., cultures were again incubated for 3 h at 37 °C. The number of ciprofloxacin-resistance mutants was determined by plating the entire culture on Nutrient agar No.2. plates containing a selective concentration of antibiotic. The total number of viable cells was determined by plating an appropriate dilution of cultures on nonselective medium. The colony forming units was counted after 24 h incubation and resistant colonies on selective plates after 72 h.

Effect of sorghum extract on the mutation frequency leading to ciprofloxacine resistance was epressed as percentage of resistance index following the formula % RI = $[(X_1/X_2)100]$ where X_1 represents resistance index of spontanes revertants and X_2 is resistance index of sorghum extract. Resistance index represents number of resistant cells divided by the number of viable cells.

Statistical Analysis

Data shown represent the mean of three independent experiments and each experiment was made in five parallel determinations. Statistical differences at $P < 0,05$ and $0,01$ were considered to be significant.

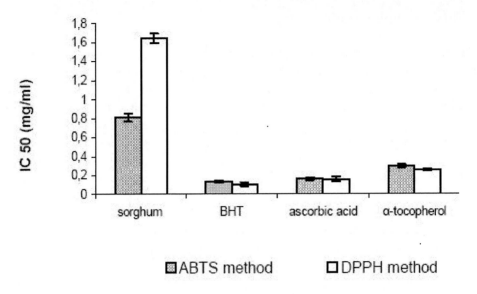

Figuare 1. Antioxidant activity of sorghum extract and synthetic standard, well-known antioxidants, measured by ABTS and DPPH test expressed as IC 50 value in mg/ml.

RESULTS AND DISCUSSION

Antioxidant Activity

The antioxidant activity of sorghum extract and synthetic antioxidant, measured by three independent methods, ABTS, DPPH, and FRAP test, is presented in Figures 1 and 2.

Antioxidant potential of sorghum extract was observed in all used methods with slight differences, but 50% of inhibition value for extract was more significantly lower when compared to commonly used synthetic antioxidants. As Figure 1 show, the best free radical scavenging ability was displayed BHT, 100 mg of sorghum extract has almost an equivalent inhibition value of 17 mg BHT for ABTS test and 6 mg BHT respectively for DPPH test. The higher antioxidant activity of sorghum determined by ABTS test than by DPPH test may be due to react ABTS cation radical with any hydroxylated aromatics independently of their real antioxidant potential. Meanwhile, DPPH does not react with flavonoids, which contain no OH-group in B-ring as well as with aromatic acids containing only one OH-group [23]. In FRAP method (Figure 2), antioxidant effect was in descending order was ascorbic acid > BHT > α-tocopherol > sorghum. The IC 50 value of sorghum extract was also higher than synthetic antioxidant but their antioxidant activity was not as significant as in previous case. The differences of IC 50 value of sorghum extract measured by FRAP assay and ABTS or DPPH test can be caused by various mechanisms of these methods. ABTS and DPPH test is indirect antioxidant methods which are based on elimination of radicals in general. In case of FRAP assay, method is based on the ability of antioxidants to reduce $Fe(3+)$ to $Fe(2+)$ [23].

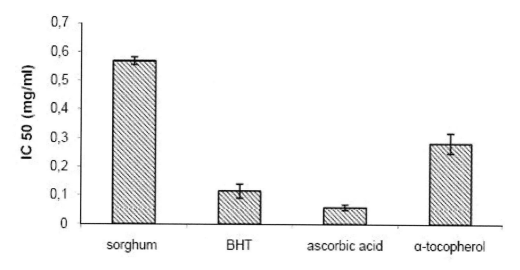

Figure 2. Antioxidant activity of sorghum extract and synthetic standard, well-known antioxidants, measured by FRAP test explained as IC 50 value in mg/ml.

Antimicrobial Activity

The results of the antimicrobial activity estimating by disk diffusion method indicated variation in the inhibitory properties of sorghum extract (Table 1, 2). In general, extract exhibited strong antimicrobial activity against gram-positive and gram-negative bacteria as well. The most significant antibacterial effect of tested extract was registered on bacterial strain *B. subtilis*, *S. aureus* and *S. typhimurium*, growth of these bacteria was inhibited in all used concentrations. The moderate variance between inhibition against gram-positive and gram-negative bacteria could be caused by differences in cell structures of bacteria. In particular, the outer membranes of gram-negative bacteria function as a preventive barrier against hydrophobic compounds [24] such as tannins which are present in many plant food including sorghum [25; 26; 9].

Table 1. The antibacterial activity of sorghum extract

sample	Dose mg/ml	Inhibition zones (cm)						
		Gram-positive bacteria			Gram-negative bacteria			
		B. subtilis	*S. aureus*	*S. epidermidis*	*S. typhimurium*	*P. aeruginosa*	*E. coli*	*E. sakazakii*
sorghum	0	NI	NI	NI	NI	NI	NI	NI
	6.25	1.0	0.7	NI	1.1	NI	NI	NI
	12.5	2.0	0.8	0.9	1.4	1.2	NI	NI
	25	1.6	0.9	1.1	1.3	1.0	1.0	1.3
	50	1.5	1.8	1.2	2.0	1.3	1.1	1.2
gentamicin	*	2.1	1.9	2.1	1.8	1.8	1.7	1.8

*20 µg/disk; NI – no inhibition.

Table 2. Antimicrobial effect of sorghum extract on growth of yeast and fungi

sample	Dose (mg/ml)	Inhibition zones (cm)						
		Yeast			Fungi			
		S. cerevisiae	C. albicans	C. maltosa	R. oryzae	A. flavus	F. culnarum	A. alternata
sorghum	0	NI	NI	NI	NI	NI	NI	NI
	6.25	NI	NI	NI	NI	NI	NI	NI
	12.5	NI	NI	NI	NI	NI	NI	1.0
	25	NI	NI	NI	NI	1.6	NI	1.5
	50	NI	NI	NI	NI	1.8	NI	1.9

NI – no inhibition.

At higher concentrations, extract also suppressed growth of fungi *A. flavus* and *A. alternata*. On the other hand, no inhibitory activity against tested yeast was observed. These antimicrobial activities could be due to content of biological components including polyphenolic compounds. Based on previous results, sorghum is significantly source of flavonoids and ferulic acid is predominant phenolic acid in extract [27].

Although polyphenols are well known for their potential biological activities, including antimicrobial at this moment we cannot say exactly, whether they are responsible for sorghum inhibitory properties. There are many researches dealing with antimicrobial effect of natural sources but there is less that could be related to sorghum. One of the few studies concerning to antimicrobial properties of sorghum extract or its variety is research published by Kil et al. [12]. Their results also indicated that extracts inhibited tested microorganisms with more or less sensitivity. According to authors, sorghum extracts, variety Jangsususu and Neulsusu showed no inhibitory effect against *B. subtilis*. On the other hand, extract from other variety was efficient only against *C. albicans*, but this effect did not confirm to us. The differences could be associated with many factors including another variety of sorghum, soil state, maturity at harvest, growing and post-harvest storage condition [28; 13].

B. subtilis as the most susceptible gram-positive and *S. typhimurium* as as the most susceptible gram-negative strain was selected for more detailed studying of the antimicrobial activity of sorghum extract by bioautography detection (Table 3).

Table 3. The bioautography detction of sorghum extract antibacterial activity

	Inhibition zones (cm)					
	Bacillus subtilis			*Salmonella typhimurium*		
Rf factor (cm)/solvent chamber	A	B	C	A	B	C
start	0.45	0.65	0.7	NI	0.45	0.40
1.0	NI	0.65	NI	NI	NI	NI
1.2	NI	NI	NI	NI	0.45	NI
1.8	NI	0.9	NI	NI	NI	NI
4.3	NI	NI	0.75	NI	NI	NI
4.4	0.95	NI	NI	NI	NI	NI
7.5	NI	0.75	NI	NI	NI	NI

Rf – retention factor; NI – no inhibition; A: ethyl acetate:hexane (2:1); B: ethyl acetate:hexane (1:2); C: toluene:ethyl acetate:formic acid (5:4:1).

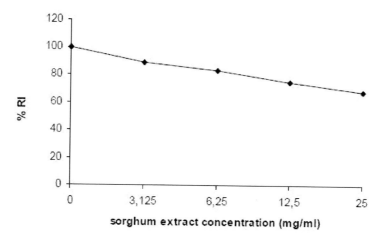

Figure 3. The effet of sorghum extract on the development of mutations leading to ciprofloxacin resistance expressed as a percentage of resistance index.

As shown in Table 3, our results suggested that tested extract consists of active components efficient on used bacterial strains. For the present, more detailed study of antimicrobial detection of sorghum extract was only preliminary. The determination of active components will be subject of further study.

Determination of Sorghum Effect on the Development of Ciprofloxacin Resistance

As shown in Figure 3, with increasing concentration of sorghum extract mutation frequency leading to ciprofloxacin resistance was slightly decreasing. The highest tested concentration of extract has decreased mutation frequency leading to 68 %. Our previous results suggested that sorghum extract effect on the development of ciprofloxacin resistance might be due to its antimutagenic activity [27]. Some researches describing positive influence of natural components such as polyphenolic compounds on bacterial growth. The decrease of oxacilin MIC in presence geraniol isolating from *Zanthoxylum pipericum* was observed [29]. According to Pillai et al. [14], the presence of green tea catechins or other antioxidants could prevent the mutations leading to resistance development by scavenging free radicals. On the other hand, the work of Birošová et al. [15] showed that some phenolic acids, including cinnamic, caffeic and ferulic acid, may increase the development of resistace. Therefore, study of natural sources ifluence on bacterial resistance is still current.

CONCLUSION

Results in this study showed that sorghum is a crop with potential biological properties, including antioxidant and antimicrobial activity against fair-sized spectrum microorganisms. Data also suggested that sorghum extract has a potential to decrease mutation frequency to resistance. However bacterial resistance is still increasingly serious problem it is important to search for new possibilities of its positive influence.

ACKNOWLEDGMENTS

This work was supported by the Slovak Grant Agency APVV (grant No. 0310-06) and by the Agency of the Ministry of Education, Science, Research and Sport of the Slovak Republic for the Structural Funds of EU as a part of the Project: "Evaluation of natural substances and their selection for prevention and treatment of lifestyle diseases (ITMS 26240220040). The authors thank to plant production station in Uhříněves for obtaining the sorghum seeds for this research project.

REFERENCES

[1] Welch, R. W. Cereal grain. In: Caballero B, Allen L, Prentice A editors. *Encyclopedia of human nutrition.* Oxford: Elsevier, 2005, 590.
[2] Dlamini, N. R., Taylor, J. R. N. and Rooney, L. W. (2007). The effect of sorghum type and processing on the antioxidant properties of African sorghum-based foods. *Food Chemistry,* 105, 1412-1419.
[3] Ragaee, S., Abdel-Aal, E.-S. M. and Noaman, M. (2006). Antioxidant activity and nutrient composition of selected cereals for food use. *Food Chemistry,* 98, 32-38.
[4] Gorinstein, S., Medina Vargas, O. J., Jaramillo, N. O., Arnao Salas, I., Martinez Ayala, A. L., Arancibia-Avila, P., Toledo, F., Katrich, E. and Trakhtenberg, S. (2007). The total polyphenols and the antioxidant potentails of some selected cereals and pseudocereals. *European Food Research and Technology,* 225, 321-328.
[5] Obilana, A. B. Sorghum: Breeding and Agronomy. In: Wrigley C, Corke H, Walker ChE editors. *Encyclopedia of grain science.* Oxford: Elsevier; 2004; 512.
[6] Salinas, I., Pró, A., Salinas, J., Sosa, E., Becerril, C. M., Cusa, M., Cervantes, M. and Gallegos, J. (2006). Compositional variation amongst sorghum hybrids: Effect of karin concentration on metabolizable energy. *Journal of Cereal Science,* 44, 342-346.
[7] Kamath, V., Chandrashekar, A. and Rajini, P. S. (2004). Antiradical properties of sorghum (*Sorghum bicolor* L. Moench) flour extracts. *Journal of Cereal Science,* 40, 283-288.
[8] Awika, J. and Rooney, L. (2004). Sorghum phytochemicals and their potential impact on human health. *Phytochemistry,* 65, 1199-1221.
[9] Dykes, L and Rooney, L. W. (2006). Sorghum and millet phenols and antioxidants. Journal of Cereal Science, 44, 236-251.
[10] Sikwese, F. E. and Duodu, K. G. (2007). Antioxidant effect of a crude phenolic extract from sorghum bran in sunflower oil in the presence of ferric ions. *Food Chemistry,* 104, 324-331.
[11] Almajano, M. P., Carbó, R., López-Jimenéz, J. A. and Gordon, M. H. (2008). Antioxidant and antimicrobial activities of tea infusion. *Food Chemistry*, 108, 55-63.

[12] Kil, H. Y., Seong, E. S., Ghimire, B. K., Chung, I.-M., Kwon, S. S., Goh, E. J., Heo, K., Kim, M. J., Lim, J. D., Lee, D. and Yu, Ch. Y. (2009). Antioxidant and antimicrobial activities of crude sorghum extract. *Food Chemistry*, 115, 1234-1239.

[13] Hussain, A. I., Anwar, F., Sherazi, S. T. and Przybylski, R. (2008). Chemical composition, antioxidant and antimicrobial activities of basil (*Ocimum basilicum*) essential oil depends on seasonal variations. *Food Chemistry*, 108, 986-995.

[14] Pillai, S. P., Pillai, Ch. A., Shankel, D. M. and Mitscher, L. A. (2001). The ability of certain antimutagenic agents to prevent development of antibiotic resistance. *Mutation Research*, 496, 61-73.

[15] Birošová, L, Mikulášová, M. and Vaverková, Š. (2007). Phenolic acid from plant foods can increase or decrease the mutation frequency to antibiotic resistance. *Journal of Agricaltural and Food Chemistry*, 55, 10183-10186.

[16] Krygier, K., Sosulsky, F. and Hogge, L. (1982). Free, esterified and insoluble-bound phenolic acids. 1. Extraction and purification procedure. *Journal of Agriculture and Food Chemistry*, 30, 330-334.

[17] Kim, K.-H., Tsao, R., Yang, R. and Cui, S. W. (2006). Phenolic acid profiles and antioxidant activities of wheat bran xtracts and zhe effect of hydrolysis conditions. *Food Chemistry*, 95, 466-473.

[18] Arts, M. J. T. J., Haenen, G. R. M. M., Voss, H. and Bast, H. (2004). Antioxidant capacity of reaction products limits the applicability of the Trolox Equivalent Antioxidant Capacity (TEAC) assay. *Food Chemistry and Toxicology*, 42, 45-49.

[19] Yen, G. Ch. and Chen, H. Y. (1995). Antioxidant activity of various tea extracts in relation to their mutagenicity. *Journal of Agriculture and Food Chemistry*, 43, 27-32.

[20] Niemeyer, H. B. and Metzler, M. (2003). Differences in the antioxidant activity of plant and mammalian lignans. *Journal of Food Engineering*, 56, 255-256.

[21] Betina, V., Baráthová, H., Fargašová, A., Frank, V., Horáková, K. and Šturdík, E. (1987*). Platňová disková metóda: Mikrobiologické laboratórne metódy*. Bratislava, ST: Alfa.

[22] Gu, L., Wu, T. and Wang, Z. (2009). TLC biouatography-guided isolation of antioxidants from fruit of *Perilla frutescens* var. *acuta*. *Food Science and Technology*, 42, 131-136.

[23] Roginsky, V. and Lissi, E. A. (2005). Review of methods to determine chain-breaking antioxidant activity in food. *Food Chemistry*, 92, 235-254.

[24] Puupponen-Pimiä, R., Nohynek, L., Meier, C., Kähkönen, M., Heinonen, M., Hopia, A. and Oksman-Caldentey, K.-M. (2001). Antimicrobial properties of phenolic compounds from berries. *Journal of Applied Microbiology*, 90, 494-507.

[25] Al-Mamary, M., Al-Habori, M., Al-Aghbari, A. and Al-Obeidi, A. (2001). *In vivo* effects of dietary sorghum tannins on rabbit digestive enzymes and mineral absorption. *Nutrition Research*, 21, 1393_1401.

[26] Taylor, J. R. N., Schober, T. J. and Bean, S. R. (2006). Novel food and non-food uses for sorghum and millets. *Journal of Cereal Science*, 44,252-271.

[27] Mošovská, S., Mikulášová, M., Brindzová, L. Valík, Ľ. and Mikušová, L. (2010). Genotoxic and antimutagenic activities of extracts from pseudocereals in the *Salmonella* mutagenicity assay. *Food and Chemical Toxicology*, 48, 1483-1487.

[28] Podsedek, A. (2007). Natural antioxidants and antioxidant capacity of Brassica vegetables: A review. *LWT-Food Science and Technology*, 40, 1-11.

[29] Hatano, T., Kusuda, M., Inada, K., Ogawa, T.-O., Shiota, S., Tsuchma, T. and Yoshida, T. (2005). Effects f tannins and related polyphenols on methicillin-resistant *Staphylococcus aureus*. *Phytochemistry,* 66, 2047-2055.

In: Sorghum: Cultivation, Varieties and Uses
Editors: Tomás D. Pereira

ISBN: 978-1-61209-688-9
©2011 Nova Science Publishers, Inc.

Chapter 8

BIOENERGY PRODUCTION FROM SORGHUM

Ana Saballos
Agronomy Department, University of Florida, U.S.A.

ABSTRACT

Sorghum bicolor is the 5th most cultivated cereal crop in the world. It has been used by humans for over 5,000 years for food, feed and fodder and it is a staple crop in arid, nutrient-poor regions of the world due to it tolerance to harsh growing conditions. Its resilient nature allows its cultivation with fewer inputs than other crops such maize, wheat and rice. Because of its reciliance and high biomass yield potential, there is a growing interest in sorghum cultivation as a bioenergy crop. Three production streams for biofuels and bio-based products can be obtained from the species: Stanch from the grain, sugar from the sweet stalks and lignocellulosic biomass from the bagasse or stover. Current sorghum varieties and hybrids were not bred for bioenergy applications, and consequently, they do not take full advantage of the genetic variation present in the species. Efforts to breed dedicated bioenergy varieties are underway, with focus in incorporating traits such as high biomass, improved sugar production, and increased stover and starch saccharification potential. This review will cover the current and proposed methodologies for bioenergy production from sorghum, and the research underway to improve the feedstock quality of the species.

1. WHY SORGHUM IS AN EXCEPTIONAL BIOENERGY CROP

Sorghum currently ranks fifth among the world's grain crops and is produced on every inhabited, with a production area of 42 million hectares. Sorghum grain is used as food primarily in drier parts of Africa, Asia and Central and South America, where the cultivation of maize, wheat, and rice is not practical. In other parts of the world, the grain is used as animal feed and as a source of starch for industrial processing, including bio-ethanol in the U.S. The vegetative parts of the plant are commonly used as forage and silage. Sweet sorghums, a special class of sorghums that tend to have tall stems that accumulate sugars, have traditionally been grown for the production of syrup and molasses. Sorghum has inherent attributes that make it particularly attractive, especially in the context of sustainable crop production in developing countries, because it has a high tolerance to abiotic stresses such as drought and heat. Under severe drought or heat, sorghum arrests its growth, but then grows rapidly when water is available or temperatures drop, thus avoiding yield losses. Sorghum employs C4 photosynthesis, which makes its water use more efficient. In addition, water loss is minimized a result of a thick layer of epicuticular waxes. Sorghum also has an extensive root system as an adaptation to arid regions with low inherent soil fertility. The root system can penetrate 1.5 to 2.5 m into the soil and extend 1 m away from the stem (Dogget, 1988). The large amount of sorghum root material contributes to the build-up of soil organic matter (SOM) after harvest of aerial plant parts. This can mitigate productivity losses due to reduction in SOM and soil erosion as a result of stover removal (Maskina et al., 1993; Wilhelm et al., 2004). Sorghum requires less fertilizer than corn to achieve high yield (Lipinsky and Kresovich, 1980) and can tolerate a wider range of soil conditions, from heavy clay soils to light sand, with a pH from 5.0 to 8.5 (Smith and Frederiksen, 2000), and dry as well as poorly drained soils. This makes it suitable for cultivation in marginal lands.

Sorghum possesses great genetic diversity for high biomass production and other useful traits. Many sorghums genotypes can grow as tall as 5 meter, have thick stalks, and large tillers. Sorghum will also grow back after removal of the stem ("ratooning"), so that two biomass harvests are possible where the climate permits. While sorghum originated in Africa, it has been grown world-wide for centuries and has adapted to many local environments. Unlike many other crops, especially biomass crops, sorghum is a diploid. This facilitates genetic studies and the introduction of mutations (Xin et al., 2009). While many candidates biomass crops are propagated vegetatively (root or stem cuttings), sorghum is propagated via seed, wich allows its easy production, storage, commercialization and planting. Sorghum is relatively easy to cross-pollinate, which permits convenient hybridization between diverse sources of germplasm, and thus enables the combination of useful traits into new cultivars. The sorghum genome has been sequenced by the Department of Energy Joint Genome Institute (JGI) (Paterson et al., 2009), and large numbers of molecular markers have been positioned on the genome sequence (Mace et al., 2009; Ramu et al., 2009; Yonemaru et al., 2009).

Because of its adaptation to suboptimal growing conditions and its potential to produce large amounts of grain, sugars, and lignocellulosic biomass, sorghum is currently receiving considerable attention as a bioenergy crop that would enable the transition from "first generation" starch-based ethanol to "second generation" cellulosic ethanol. Three production streams for energy production can be obtained from sorghum. Together with maize, sorghum grain is currently a feedstock for starch-based ethanol (Agri-Energy Solutions, 2010). Sweet sorghums represent a transition between starch-based and cellulosic ethanol, because the sugars they produce in the plant sap can be readily fermented (i.e., without expensive pretreatment) once they have been extracted from the stem in a manner very similar to the production of ethanol from sugarcane in Brazil. The transition from starch-based to cellulose-based ethanol is necessary because the use of grain for the production of fuel is considered unethical ("food versus fuel"; Tenenbaum, 2008) and unsustainable due to the high demand for nitrogen fertilizer and energy inputs to produce the grain that drives up the cost of production and contribute to greenhouse gases emissions (Eickhout et al., 2006; Kim and Dale, 2008). Cellulosic ethanol, in contrast, is not produced from grain, but from lignocellulosic biomass. The bagasse (the crushed stems remaining after juice extraction) from sweet sorghum or the lignocellulosic biomass from high-biomass non-sweet types can be used as a feedstock for cellulosic ethanol (Vermerris et al., 2007). Cellulosic ethanol has a better energy balance than starch-based ethanol (Hill at al. 2006) and dramatically decreases the emission of greenhouse gases (Farrell et al., 2006). Dedicated biomass crops are required to achieve the level of production necessary to significantly displace petroleum-based fuels. High biomass sorghums have the potential to become a leading component in the biomass species portfolio of the future.

2. GRAIN SORGHUM FOR STARCH-BASED ETHANOL

The logistics (harvest, transport, storage) of grain sorghum are well developed and similar to maize. Ethanol plants located in the drier areas of the US frequently use grain sorghum together with maize. The advantages of the use of grain sorghum for ethanol production include (Agri-Energy Solutions, 2010):

1. The inputs for producing grain sorghum are lower than those of maize, which in turn translates in lower feedstock cost at the processing plant.
2. The by-product of ethanol production (distillers' grains) have higher protein rates than those from maize.
3. Grain sorghum produces the same amount of ethanol per bushel as maize.

About 30% of the sorghum grain crop is processed for ethanol (Agri-Energy Solutions, 2010). This represents only about 2.5% of the total grain used for ethanol production, despite the above mentioned advantages. Market misconceptions and inconsistencies of supply have limited its use. It seems then that the most effective strategy to increase the share of sorghum grain as feedstock for ethanol production is the dissemination of the correct information about grain sorghum to the ethanol producers. It is also necessary to increase the marketing efforts of grain sorghum to the ethanol plants and to inform producers of the potential market represented by the ethanol plants.

2.1. Energy Production from Starch

Starch-based ethanol production is currently the only biofuel production system that is considered a mature technology. In dry milling (Fig. 1), the kernel is ground to produce the "meal" and expose the starch chains. It is then mixed with water. Enzymes are added to the slurry (mash) to reduce the viscosity of the mixture (liquefaction) and to cut the starch chains into the glucose units (saccharification). After the enzymatic digestion, the mash is cooked to reduce the bacterial population. It is then cooled and fermentative microorganisms (usually baker's yeast, *Saccharomyces cerevisiae*) are added. After the fermentation process is completed, the "beer" is transferred to a distillation column. Ethanol is distilled to a maximum of 96%. To achieve higher purity, the mixture is dehydrated using a molecular sieve. The anhydrous ethanol is mixed with 5% denaturant (such as kerosine) to make it unfit for human consumption (Nichols and Bothast 2008). The "stillage" (the remaining slurry after distillation) is separated into its solids and liquid components, dried, and sold as animal feed (dried distiller's grains with solutes, DDGS).

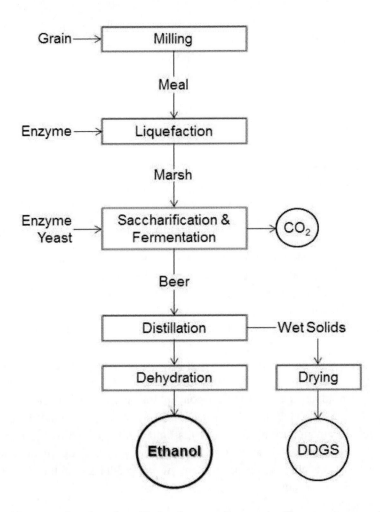

Figure 1. Dry mill process for ethanol production from sorghum grains. Process steps are shown in boxes and products in circles. Open text represents the inputs for each step of the process. Adapted from Nichols and Bothast, 2008.

2.2. Improvement of Grain Sorghum for Bioenergy Applications

As with any starch feedstock, the primary targets for improving bioethanol sorghum are greater yield of grain per hectare and increased yield of net energy per weight of grain. Improving the grain yield per hectare has been the main target of sorghum breeding for all other applications, and it is reviewed elsewhere. The increase in net energy output can be achieved either by reducing the energy needed to produce the ethanol or increasing the energy obtained from the same amount of feedstock. Both measures can be improved by tailoring the chemical composition of the grain to the process employed for the production of fuel. In the following section, the compositional factors that influence the ethanol potential of the sorghum grain will be discussed.

2.2.1. Compositional Factors Impacting Ethanol Production

The sorghum grain (commonly referred as kernel) is composed of three parts: Pericarp, embryo, and endosperm (Doggett, 1988). Pericarp is the protective cover of the seed. The cells of the outer portion of the pericarp are thick-walled and generally covered by a thin layer of wax. A testa layer is located under the pericarp. Both the pericarp and testa can be colored or colorless, depending on the genetic composition of the female plant. The endosperm is the main component of the sorghum grain. It is a triploid tissue, therefore its appearance and composition will depend on both the male and female parental plant genotypes. The vast majority of the starch granules are found in the endosperm. Sorghum endosperm has a corneous and a floury area. The starch grains in the corneous endosperm are surrounded by a continuous protein matrix, which hinder the accessibility of the starch to hydrolytic enzymes and water. The floury endosperm has a discontinuous protein matrix, and loosely packaged starch granules. The embryo is the result of the union one male and one female gamete. It contains the majority of the oil present in the seed. The relative proportions of the seeds parts and its components vary among sorghum genotypes.

Genotypic and environmental factors influence the composition of cereal grains. Accordingly, sorghum lines have been shown to differ in their ethanol production potential (Zhan et al., 2003; Zhao, 2009). High starch content, floury endosperm, and the presence of the *waxy* (*wx*) mutations are characteristics positively associated with ethanol production (Zhan et al., 2003; Wu et al., 2007; Wong et al., 2010). High-tannin sorghum lines, such as the bird-resistant varieties, have significant lower conversion efficiency. The higher tannin content had a negative effect on the liquefaction of the mash (Wu et al., 2007), likely due to the binding and inactivation of the hydrolytic enzymes by tannins. The protein digestibility, but not the protein content, is correlated to the starch conversion efficiency. Protein digestibility is correlated with the degree of cross-linking of the protein matrix, which in turn can influence the degree of accessibility of the starch in the granules to the hydrolytic enzymes (Wu et al., 2007).

2.2.2. Target Traits for Grain Sorghum Breeding for Bioenergy Purposes

Breeding for bioenergy traits in grain sorghum has received little attention compared to biomass and sweet sorghum. Fortunately, many of the traits that confer increased digestibility for human food and animal feed are the same that increase the conversion efficiency for ethanol production. The information obtained from nutritional research on the compositional traits and their genetic basis can be readily used to improve sorghum's conversion efficiency for ethanol production.

2.2.2.1. Waxy Sorghums

The starch in the endosperm of cereal grains consist of two types of glucose polymers: a mostly linear form called amylose, and a highly branched form called amylopectin (James et al., 2003). *Waxy* (*wx*) mutants have altered starch composition with reduced amylose content. Grains from *wx* sorghum have altered characteristics, including higher ethanol conversion rates and higher protein digestibility, compared to wild type sorghum (Rooney and

Pflugfelder, 1986; Sang et al., 2008, Wong et al., 2009). Two alleles of the *wx* gene have been cloned (Sattler et al., 2009). The mutations are located in the *granule-bound starch synthase* (*GBSS*) gene (Sb10g002140). The wx^a allele has a large insertion in the gene, resulting in no accumulation of the GBSS enzyme in the starch granule. The GBSS activity on starch granules from wx^b genotypes was less than 24% of that of wild type genotypes. Molecular markers able to differentiate between both mutant alleles and the wild type enzyme were developed (Sattler et al., 2009). These markers can facilitate the breeding efforts to take advantage of the good processing characteristics of the *wx* mutations.

2.2.2.2. Tannins

Tannins are polyphenol compounds that are associated with poor protein digestibility in sorghum (Doggett, 1988). They are found in higher concentrations in brown pericarp and dark testa types, although white pericarp types with dark testa can contain high levels of tannin. Condensed tannins are found mainly in the testa layer. The presence of the dominant form of two genes, designated B1 and B2, are necessary for the accumulation of condensed tannins in the testa (Stephens, 1924). Recently, two quantitative trait loci (QTL) for tannin content were identified (Xiang, 2009). The QTL and their interaction explain 74% of the phonotypic variation in a population of recombinant inbred lines (RILs) from a cross between low and high tannin sorghum lines. The QTL are located in the distal portion of long arm of chromosomes 2 and 4. While no candidate genes underlying the QTL were identified in this study, the regions identified provide the base for high-resolution mapping. The markers linked to the high/low tannin content can be directly deployed in breeding programs. Most of the sorghum produced in U.S.A. is non-tannin sorghum (Awika and Rooney, 2004), but many of the landraces and wild relatives, which represent an important source of genetic variability, are tannin sorghums. The availability of molecular markers for tannin content will facilitate the use of the exotic germplasm in grain sorghum breeding.

2.2.2.3. Protein Digestibility

The major protein fractions in sorghum grains are kafirins and glutelins (Awika and Rooney, 2004). Kafirins are seed storage proteins, found in protein bodies tightly bound around starch grains. They are divided into four subclasses, α-, β-, γ- and δ-kafirin. β- and γ-kafirin are cysteine-rich proteins that form molecular disulfide bonds, contributing to decreased protein digestibility. Transgenic down-regulation γ-kafirin has been proposed as a way to improve protein digestibility (Wong et al., 2010). The location of the genes encoding the kafirin proteins has been identified *in silico* (Laidlaw et al., 2010) using the genome sequence (Paterson et al., 2009) and the allelic diversity of the genes from a subset of diverse entries was described. Among the alleles identified, a β-kafirin null mutant was found. The influence of the alleles on ethanol production characteristics of the grain remains to be investigated, but it is likely that some of the alleles may confer improved processing quality to the sorghum grain. For example, the flour of lines containing the null β-kafirin allele have lower viscosity compared to lines containing a functional allele, a characteristic likely to be advantageous for bioprocessing.

2.2.2.4. Grain Mold Resistance

Resistance to grain molds is not directly related to ethanol production, but still an important consideration. The DDGS, a by-product of processing grain to ethanol, is sold to livestock producers for use in the feed of cattle and swine. The sale of this by-product contributes to the profitability of the ethanol plant operation (Mathews and McConnell, 2009). *Fusarium* and *Aspergillus* species predominate within the sorghum grain mold complex, and some of their members are able to produce mycotoxins, such as fumosine and aflatoxin (Leslie et al., 2005). The presence of mycotoxins in the grain has negative health effects and reduces the weight gain of the livestock (Wu and Muckvold, 2008), therefore reducing the value of distillers' grains for the livestock operations. Unfortunately, traits that have been shown to confer resistance to grain mold are also negatively related to ethanol production efficiency. In particular, vitrous endosperm and high-tannin sorghums have higher resistance (Esela et al., 1993; Audilakshmi et al., 1999), although some white-seeded mold-resistant genotypes have been reported (Prom and Erpelding, 2009) suggesting than additional factors beside tannins may be able to confer high levels of resistance. Some other traits associated with moderated levels of resistant include dark colored glumes (Audilakshmi et al., 2009) and the presence of antifungal proteins in the seed (Rodríguez-Herrera et al., 1999).

2.3. CURRENT EFFORTS TO IMPROVE THE USE AND PROCESSING OF GRAIN SORGHUM FOR BIOFUEL

The National Sorghum Growers Association has identified the biofuel industry as one of the major possible expansion areas for the grain sorghum market (Agri-Energy Solutions, 2010). Consumer misconceptions regarding grain sorghum quality as feedstock for ethanol plants was one of the main limitations to its widespread use for this purpose. Maize feedstocks are routinely monitored to determine quality and price at the ethanol plant. A similar system based on standardized methods to determine grain quality are needed to improve consumer confidence in grain sorghum. Studies such as that of Wu et al., (2007) have identified specific characteristics that affect ethanol production from grain sorghum. The development of protocols that are able to measure those traits to predict the ethanol yield of grain sorghum in a quick and cost efficient manner would benefit both marketing and breeding efforts of grain sorghum for bioenergy.

3. BIOMASS SORGHUMS

3.1. What are Biomass Sorghums?

While the majority of the sorghum in the US is cultivated for its grain and it has been bred to have short stature and a high panicle/stem ratio, sorghum is a versatile species with variants that can produce additional products. Non-grain varieties, including sweet, forage, and exotic material can accumulate high amount of vegetative biomass (Fig. 2). Traditionally, sorghum varieties with thin stems and leafy plant architecture have been grown for forage (forage sorghums and sudangrass) and varieties that accumulate sucrose and glucose in the stalk ("sweet sorghums") have been grown for syrup production. For the purpose of this review, sorghums whose main product is other than grain will be referred as biomass sorghum. Both the lignocellulosic biomass and the stem sugars can be used as feedstocks for bioenergy production.

Figure 2. High biomass sorghum germplasm from the University of Florida sorghum breeding program.

3.2. Energy from Stem Sugars

Sweet-stem varieties have high concentrations of soluble sugars (mostly sucrose, glucose, and fructose) in the plant sap. Fermentative microbes such as yeast and certain strains of *E. coli* can directly use these sugars for the production of alcohols and acids. Sorghum sugars can also be used to grow heterotrophic algae. The lipids in the algae's cells are then processed for the production of biodiesel (Gao et al., 2010). The easy accessibility of readily fermentable sugars combined with high yield of green biomass make sweet sorghums a promising candidate for a dedicated bioenergy crop.

3.2.1. Processing Methods

Although the concept of the use of sweet sorghum for sugar and ethanol production was proposed at least 30 years ago (Nathan, 1978), no single technology has yet emerged as the most favorable alternative for its processing. Several methods have been proposed in the literature (Almodares and Hadi, 2009). Most commonly, the sugarcane model is followed. In this model the saccharine juice is the main product obtained to be fed into the ethanol production process. The stalks are harvested with a cane harvester (Fig. 3), the panicles removed, and then the stems are transported to a mill where they are crushed, allowing extraction of the juice. The juice coming out of milling can be sterilized (Quintero et al., 2008) or fed directly into the fermentation tank (Chohnan et al., 2011). The sugars from the pasteurized or unpasteurized juice are then fermented by yeast to produce ethanol. After extraction of the juice, the crushed stems (bagasse) can also be used as lignocellulosic feedstock for ethanol production or burned for thermal energy (Fig. 4). The grains from sweet sorghum can be utilized in the same manner as grain sorghum (Reddy, et al., 2007), for food, feed or starch ethanol production. Plants designed to distill ethanol from sugarcane juice to can easily use sweet sorghum as feedstock, extending the production period of the processing plant (Kim, 2010).

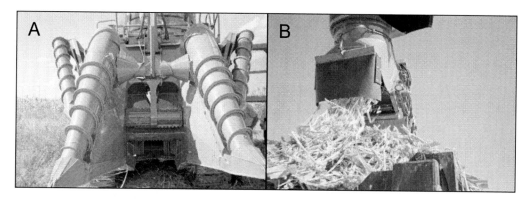

Figure 3. Sweet sorghum harvesting. A. The John Deere 3520 cane harvester, developed for sugarcane, has been successfully used for harvesting sweet sorghum in Florida. B. The product of the cane harvester, 4 to 6 inches billets of sweet sorghum stems, is collected in a trailing wagon. Photographs courtesy of Steve Segrest.

Beside the sugarcane model for sweet sorghum processing, other systems have been proposed. In-field juice extraction and fermentation systems and solid state fermentation of chopped stalks are all possible (Bryan et al., 1985; Gibbons et al., 1986; Li et al., 2004; Kundiyana et al., 2006). Compared to harvesting with a cane harvester, harvesting sweet sorghum with a forage chopper results in better biomass density and more efficient transportation of the stalks to the processing plant, although sugar concentration in the chopped stalks showed a more rapid decline compared to the stalks harvested whole (Keating et al., 2004). Beside ethanol, production of hydrogen from sweet sorghum juice has been explored (Claassen et al., 2004; Antonopoulou et al., 2007).

A common challenge for all the processing strategies to produce fuel from sweet sorghum is the potential loss of stem sugars that can occur within a few days after harvesting. Depending on the harvest system and storage conditions, losses of up to 49% of the fermentable sugars are reported for chopped stems stored for 7 days at 30°C, while insignificant losses were observed for whole and billet-cut stems in the same conditions (Eiland et al., 1983). Loses of 3 to 21% of total sugar content in whole stems stored under polyethylene film were observed after 30 days of storage (Mei et al., 2008). Due to bacterial contamination, the deterioration of the quality of the unrefrigerated extracted juice is severe, with a 20 to 100% loss of fermentable sugars within three days of storage (Bellmer et al.,

2008; Wu et al., 2010). Several methods have been proposed to solve this problem. By staggering planting dates and carefully selecting varieties with different maturity dates, the harvest period can be fine tuned to the processing capacity of the plant, circumventing the need for extended storage. In-field harvest of the juice and in-farm fermentation systems have been proposed (Bele, 2007; Kundiyana et al., 2006). Long-term preservation of chopped stems have been achieved by ensilaging in the presence of formic acid (Schmidt et al., 1997) and by the addition of sulfur dioxide (Eckhoff et al, 1984). The extracted juice can be concentrated by evaporation to over 50% soluble solids, allowing storage in the form of syrup. The syrup needs to be diluted by the addition of water before it can be fermented.

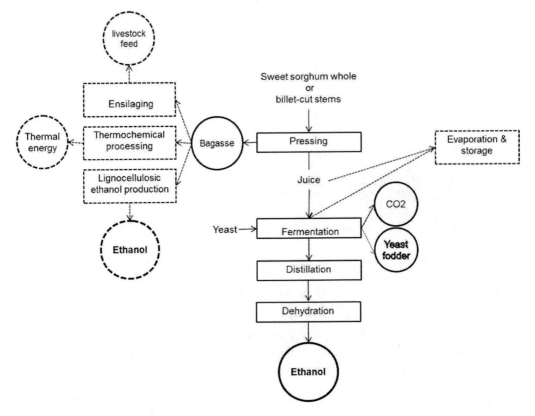

Figure 4. Processing of sweet sorghum biomass. Process steps are shown in boxes and products in circles. Open text represents the inputs for each step of the process. Pretreatment inputs,and steps and products in dashed lines will vary depending in the choice of technology.

3.3. Energy from Lignocellulosic Biomass

Sorghum can be an abundant source of lignocellulosic material, both as agricultural residue (the stover remaining in the field after the harvest of the grain) or as a dedicated energy crop. Typically, the grain represents only about 30 to 40% of the above-ground biomass of the sorghum plant (Prihar and Stewart, 1991; Patil, 2007). Much of the solar energy captured through photosynthesis is stored in the vegetative parts of the plant, namely leaves and stems. In the grain, the energy is stored in the form of starch, which is relatively easy to process for fuel production, while most of the energy in the vegetative parts is found in the form of lignocellulosic tissue. Cells in the plants' organs are protected by a wall formed mostly of cellulose, hemicellulose, and lignin (Carpita and McCann, 2000). Cellulose is the main component of the cell wall and the most common organic compound on earth. As with starch, it is also composed of glucose molecules, but cellulose and starch differ in their linkage types and organization. Cellulose is organized as a para-crystalline array of several dozen linear (1→4)-β-D-glucan chains (Delmer, 1999). The cellulose chain can be hydrolyzed to glucose units for fermentation to ethanol, but more processing is required to release the sugar monomers due to the nature of the linkage between the units and their crystalline structure (Gardner et al., 1974). In addition, cellulose microfibrils are embedded in a matrix of hemicellulosic polysaccharides and lignin (Carpita and Gibeaut, 1993). Lignin hinders the hydrolysis of cellulose by reducing the accessibility of the cellulose to the hydrolytic enzymes and by irreversible binding to the enzymes, causing their inactivation (Chernoglazov et al., 1988; Converse, 1993; Houtman and Atalla, 1995; Mooney et al., 1998; Chang and Holtzapple, 2000; Ramos, 2003; Palonen, 2004).

3.3.1. Processing Methods

Lignocellulosic materials such as sorghum bagasse and stover are low-cost, abundant, and renewable resources. Currently there is intense research in technologies to convert them into fuels for transportation and/or electric power in a cost-effective way. Broadly speaking, conversion technologies can be divided into thermochemical or biochemical processes. The characteristics considered advantageous in the feedstock material will depend on the conversion process used.

3.3.1.1. Thermochemical Processes

Co-firing (burning) of biomass for thermal and thermo-electric power generation has been investigated (Botha and Blottnitz, 2006; Basu et al., 2011). It has been successfully implemented in conjunction with fermentative ethanol production to provide the energy needed to distill the ethanol (Dias et al., 2010). In gasification, the biomass is broken down to synthesis gas ("syngas") using heat. Syngas is composed primarily of carbon monoxide and hydrogen. Syngas can then be catalytically converted to ethanol or used to generate hydrogen. Pyrolysis involves the deconstruction of the biomass by heat in the absence of oxygen, producing a bio-oil that can be further refined to transportation hydrocarbons or used in the production of hydrogen (Schwietzke et al., 2008).

3.3.1.2. Biochemical Processes

The production of ethanol from lignocellulosic biomass involves a chemical, thermal, or thermochemical pretreatment (Mosier et al., 2005) to disrupt the rigid structure of the cell wall, followed by incubation with cellulases, a consortium of enzymes that hydrolyze ("saccharify") cellulose into monomeric β-D-glucose. The glucose can be readily fermented to ethanol by microorganisms such as yeast or bacteria. The ethanol produced is then distilled and dehydrated, as with starch- and sugar-based ethanol technologies. The residue from the fermentation tank contains mostly lignin and microbial biomass and can be used for the production of high-protein animal feed, bio-materials, and/or thermal power generation (Fig. 5). A review of proposed mature process systems is found in Laser et al., (2009). Based on techno-economic analyses, the production of cellulosic ethanol is considerably less energy intensive and its production generates fewer greenhouse gases compared to either starch-based ethanol or gasoline (Farrell et al., 2006; Ragauskas et al., 2006; Schmer et al., 2008).

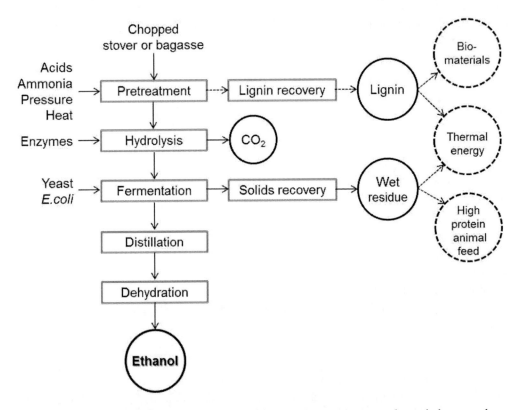

Figure 5. Biochemical processing of lignocellulosic biomass. Process steps are shown in boxes and products in circles. Open text represents the inputs for each step of the process. Pretreatment inputs, and steps and products in dashed lines will vary depending in the choice of technology. Adapted from Lu and Mosier, (2008) and Laser et al., (2009).

The management of sorghum biomass material prior to delivery to the processing plant presents common challenges independent of the choice of bioprocessing technology. Many of the high-biomass sorghums are non-senescent varieties, which results in high moisture content (70-80%) at harvest. To reduce transportation costs and allow storage of the material, the moisture content needs to be reduced to 15-20%. In drier environments in temperate locations, the crop can be left in the field after exposure to a killing frost. The use of herbicides to terminate the growth and force senescence is being investigated (Blade Energy Crops, 2010). In-field drying of biomass conditioned with a forage windrower can be used to further reduce moisture if the climate of the area allows it (Rocateli et al., 2009). Stored biomass can suffer deterioration due to adverse weather conditions. Systems that allow immediate processing of the harvested material reduce this risk. Staggered planting and harvest to continually supply material to the biorefinery would represent an ideal solution, but it is not a feasible option in areas without a year-around growing season. Most commonly, harvested biomass is compacted into bales for storage.

3.4. Biomass Sorghums Genotypes

3.4.1. Sweet Sorghum

Sweet sorghums have traditionally been produced in Louisiana, Texas, and Kentucky for syrup production. Older sweet sorghum varieties were developed from the mid-1800s to the middle of last century from early Chinese and African introductions and later collections from Africa and Australia. The main objectives in the early breeding programs were the production of syrup and later of crystalline sugar for human consumption. The USDA had an active sweet sorghum breeding program in Meridian, MS, which was closed in 1983. Improved lines with high sugar content and disease resistance were developed in that program. The germplasm material is now located at the USDA-ARS, Plant Genetic Resources Conservation Unit (PGRCU), Griffin, Georgia, and can be requested from the GRIN database (http://www.ars-grin.gov/npgs). Many of the sweet sorghums currently being evaluated were not selected with ethanol production in mind.

Table 1. Reported performance of various sweet sorghum and sweet-stemmed sudangrass varieties. [1]Study conducted in Schriever, LA. The lines where harvest 138 days after planting (DAP). [2]Study conducted in Iran. [3]Study conducted in Iran, under salinity stress of 2 deciSiemens/meters. [4]Yield of sugars calculated based on the reported percentage of sugars in the stem assuming 50% of juice extraction. [5]A multi-location study conducted in Nebraska. The range shows the higher and lower value observed. A dash indicates values not reported in the studies.

	Variety	Fresh weight Ton ha^{-1}	Brix % solubles	Hexose sugars Ton ha^{-1}	Lignocellulosic biomass Ton ha^{-1}	Hexose sugars Ethanol yield (L ha^{-1})	Lignocellulosic biomass Ethanol yield (L ha^{-1})	Total
Almodares and Sepahi 1996[2]								
	Brandes	77.1	18.7	-	-	-	-	-
	M81-E	103.6	16.0	-	-	-	-	-
	Rio	95.0	22.4	-	-	-	-	-
	Theis	100.1	19.1	-	-	-	-	-
Almodares et al. 2008[3]								
	Keller	67.4	-	10.05[4]	-	-	-	-
	Sofra	56.0	-	7.73[4]	-	-	-	-
	Kimia	35.2	-	3.67[4]	-	-	-	-
Tew et al. 2008[1]								
	MMR 333/27	67.4	-	4.6	15.6	2730	5900	8620
	MMR 333/47	105.3	-	8.3	23.4	4980	8860	13840
	Dale	81.2	-	8.4	15.3	4980	5810	10800
	M81-E	89.1	-	9.5	15.5	5680	5890	11570
	Rio	63.9	-	6.5	10.0	3920	3780	7700
	Theis	89.6	-	10.1	16.6	6060	6270	12340
	Topper	85.2	-	9.6	13.2	5780	5000	10770
Wortmann et al. 2010[5]								
	M81-E	27.57-69.9	-	1.45-5.31	-	967-3530	-	-
	Keller	34.24-59.45	-	2.27-3.33	-	1509-2212	-	-

There is a wide variation in the Brix value (1° Brix represents 1 g soluble solids in a volume of 100 ml of solvent) between sorghum cultivars. A sorghum variety is usually considered a "sweet sorghum" if its Brix value exceeds 14°. High values of up to 23° Brix have been observed, but most commonly the reported range is 16-20° Brix (Almodares and Hadi, 2009; Audilakshmi et al., 2010; Pfeiffer, 2010). Green biomass yields vary with location and variety from 20–120 ton/ha of fresh weight, with the percentage of extractable juice ranging between 30 and 50% (Channappagoudar et al., 2007; Tew et al., 2008). The main non-structural carbohydrate (NSC) in the stem is sucrose (~89%) followed by glucose and other simple sugars (~8%) and starch (~3%) (Sherwood, 1923). Traditional sweet sorghum varieties have low grain yield, but the interest in double purpose types (grain for food and fodder, sugars for ethanol production) have prompted the development of varieties with more balanced grain/sugar production in China and India (Li et al., 2004; Reddy et al., 2007). Table 1 shows the yield of several sweet sorghum varieties in controlled studies.

Table 2. Reported performance of various forage and high-biomass sorghum varieties. DM: Dry matter. [1]Study conducted in Texas. [2]Study conducted in Italy under well-watered conditions. [3]Study conducted in Italy under variable water supply. That range shown represent the yield under well-watered and water-limited conditions. [4]Study conducted in Italy. [5]Study conducted at El Reno, OK. [6]Study conducted in Italy.

	Variety	DM yield Ton ha^{-1}
McBee et al., 1988[1]		
	M35-1	16.9
Dolciotti et al., 1998[2]		
	H 173	20
Habyarimana et al., 2004a[3]		
	H 132	71-43
	Abetone	57-24
Habyarimana et al., 2004b[4]		
	H 132	27.7
	H 128	19.6
	Abetone	23.7
Venuto and Kindiger, 2008[5]		
	Piper	16.1
	Ultrasorgo	33.1
	Tentaka	40.3
Quaranta et al., 2010[6]		
	H 128	36.4
	H 133	38.7
	H 132	42.7

3.4.2. Lignocellulosic Biomass Sorghum

Many exotic and forage genotypes are photoperiod-sensitive. Photoperiod-sensitive sorghum genotypes require a short day length of less than 13.2 to 12.2 hours to trigger the transition from the vegetative to the reproductive stage of grow (Miller et al., 1968, Rooney and Aydin, 1999). When grown in non-tropical latitudes, they can reach heights of up to 5 m and accumulate high amounts of vegetative biomass due to the extended duration of their vegetative stage. Many exotic and forage genotypes are photoperiod-sensitive. Forage sorghums, sudangrasses, and tropical material were the first candidates for lignocellulosic biomass sorghum, and several studies have investigated their production potential (Table 2). Some of the quality factors that are important for forage production may not be relevant for bioenergy sorghums, however (Rooney et al., 2007). For example, nutritional value and palatability of forage decreases as the plant ages, while the amount of biomass increases. Because of this, sorghums for forage applications are harvested before they reach maximum biomass accumulation potential. As a result, forage sorghum varieties are not usually bred for late season standability. Resistance to lodging at high biomass levels would require thicker and woodier stems, a negative trait for forage quality. With the recent interest in lignocellulosic energy production, dedicated sorghum lines for biomass production are being investigated in several research institutions. The bioenergy seed company Ceres is currently marketing the first bioenergy hybrid varieties under the brand name "Blade® Energy Crops".

Lignocellulosic biomass from sorghum is composed of 24-38% cellulose, 12-22% hemocellulose, and 13-17% lignin (Amaducci et al., 2004). The cellulose and hemicellulose fractions are the feedstock for biochemical (fermentative) ethanol production. As discussed above, lignin is detrimental to this process. However, lignin has a higher heating value than cellulose (24.1 and 11.7 MJ kg^{-1}, respectively; Raveendra and Ganesh, 1996); therefore, in thermochemical processes the lignin fraction can contribute to increased energy output.

4. AGRONOMIC CONSIDERATIONS

The diversity of sorghum genotypes and the broad variety of climates and soil conditions in which sorghum can be cultivated result in large differences in sorghum production practices. The general considerations regarding agronomic practices for grain and forage sorghum will apply to biomass sorghum (Table 3). Sorghum is drought tolerant and can grow in a wide range of soils and soil conditions, but optimal yields still require the appropriate addition of nutrients and water. Therefore, it is critical to understand the differences in yield potential between the sorghum varieties, the characteristics of the site, and the production goals of every operation.

The field management of sweet sorghum for bioenergy production is likely to be similar to the recommended management of sweet sorghum for syrup production. Detailed information on cultural practices for sweet sorghum can be found in Mask and Morris, 1991, and Blitzer, 1997. As a bioenergy crop, sorghum needs to be able to produce high amounts of biomass with a minimum level of inputs in order to be economically viable. Several studies have examined the feasibility of producing sweet sorghum under various stresses, such as water-limited environments, saline soils, and nitrogen limitations (refer to tables 1 and 2). As expected, stress environments reduce the yield of sweet sorghum, but, in general, the level of production is still economically feasible. Yields of 24.7 to 103.7 ton ha^{-1} fresh weight have been obtained on phosphate-mined land in central Florida (Steve Segrest, *per. comm.*). Large variation in biomass yield was observed depending on the variety and fertilization regimen used. Considerable variation exists in the performance of sweet sorghum lines in different environments, so selection of varieties adapted to the intended production zone is of great importance.

The concentration of sugars in sweet sorghum juice reaches its maximum level between soft dough and physiological maturity of the grain. In large-scale operations, planting varieties with different growing cycle lengths (different days to maturity) and/or staggered planting should be considered to facilitate the harvest of the material at its peak and to allow continuous supply to the biorefinery.

Agronomic recommendations for lignocellulosic biomass sorghum are similar to those of forage sorghums. A detailed guide for forage sorghum in the Southeastern U.S. is available from the Florida Cooperative Extension Service (Newman et al., 2010). In 2010, Blade® Energy Crops released a management guide specifically tailored to high-biomass sorghums hybrids (Blade® Energy Crops, 2010).

There is ongoing research on the best management practices for biomass sorghum in a variety of environments (e.g., Rajendran et al., 2000; Tsuchihashi 2004; Amaducci et al., 2004; Sakellariou-Makrantonaki et al., 2007; Teetor et al., 2010; Wortmann et al., 2010), therefore agronomic recommendations are continuously being updated. However, sources of biomass sorghum seed, harvest and storage methods and harvest equipment are areas in which producers are likely to find a lack of information. Biomass sorghum for energy production is an industry on its infancy, and there is ongoing research on the best management practices for biomass sorghum in a variety of environments (e.g., Rajendran et al., 2000; Tsuchihashi, 2004; Amaducci et al., 2004; Sakellariou-Makrantonaki et al., 2007; Xiong et al., 2008; Teetor et al., 2010; Wortmann et al., 2010); therefore, agronomic recommendations are continuously being updated.

Sources of biomass sorghum seed, harvest and storage methods, and harvest equipment are areas in which producers are likely to find a lack of information. The future of the bioenergy industry depends in many unknown economic and political factors. Because of that, seed and agro-equipment companies have only minimally invested in products for the bioenergy market. Currently, the only dedicated biomass hybrid sorghum seed source marketed by Blade® Energy Crops. Sorghum x sorghum (ES 5200 and ES 5201) and sorghum x sudangrass (ES 5140 and ES 5150) hybrids for lignocellulosic production are available from them. Most other seed companies are developing hybrids that can perform both as forage and biomass sorghums. Syngenta is offering 350FS, a hybrid sweet sorghum for forage and bioenergy applications. Advanta seeds and Mendel Biotechnology (in collaboration with MMR genetics) have announced plans to develop bioenergy sorghum hybrids, but no commercial seed is available to date. Research institutions in the U.S., including the University of Nebraska, Texas A and M, and the University of Florida, have active germplasm development programs. In India, the International Crops Research Institute for the Semi-Arid Tropics (ICRISAT) has released several pure-line and hybrid varieties for double-purpose (grain and sugar) production.

Table 3. General recommendations for biomass sorghum. Specific recommendations will vary according to the area of production. Recommendation for forage and sweet sorghum from local extension offices can be used as a guideline. Information adapted from Mask and Morris, (1991), Saballos, (2008), Newman et al., (2010), and Blade Energy Crops, (2010)

Planting time	When the soil temperature (at a depth of 5 cm) reaches 21°C or above and the daylength exceeds 12 hours and 30 min.
Seed treatment	Herbicide safener: Concep®III Fungicides: Captan, Maxim®, Thiram®, Apron® Insecticides: Lorsban® Systemic insecticides: Cruiser®, Gaucho®, Poncho®
Soils	Tolerant to a wide range of soil conditions. pH range of 5.8 to 8.5. Sweet sorghum is reported to be sensitive to acid soils. Avoid water-logged soils.
Land preparation	Biomass sorghums: No-till, Conservation tillage, Strip till
	Sweet sorghum: Conventional till, no-till, conservation tillage, strip till
Nutrients	Depletion of nutrients due to green biomass removal should be monitored. The recommendations presented here are based on a medium-fertility soil test level.
	Nitrogen: Lignocellulosic biomass sorghum: 4.5 kg of nitrogen per dry ton of biomass removed. Sweet sorghum: 45 kg ha^{-1} of nitrogen is recommended. Excess nitrogen reduces sugar concentration in the juice.
	Phosphorus and Potassium: Lignocellulosic biomass sorghum: 73 and 135 kg ha^{-1} of P_2O_5 and K_2O, respectively. Sweet sorghum: 45 kg ha^{-1} of each P_2O_5 and K_2O.
Planting method	Lignocellulosic biomass sorghum: row planting with row-crop planter or drill. Broadcast planting is also possible for forage sorghum.
	Sweet sorghum: row planting with row-crop planter or drill.
Seed rate and spacing	Lignocellulosic biomass sorghum: 300000 to 185000 seeds ha^{-1}.
	Sweet sorghum: 60 to 100 cm between rows, 9 to 12 plants m^{-1}.
Weed control	Under high weed pressure, a pre-plant burn-down with broad spectrum herbicide followed by a pre-emergence application is recomended. Both pre- and post-emergence herbicides are available for sorghum. Examples of herbicides labeled for sorghum include s-metolachlor (pre-emergence) and atrazine (pre- and post-emergence).
Water requirements	Lignocellulosic biomass sorghum: 760 mm of rainfall per year. Sweet sorghum: 400 mm of water. Sufficient water during seedling establishment and during the phase of high biomass accumulation (80-90 days after emergence) is necessary to achieve the yield potential of the variety.

The harvest, transportation, and storage of energy crops present unique challenges due to the high volume and perishable nature of the biomass. So far, the sugarcane and forage models have been followed for biomass sorghum, but sweet sorghum harvesting differs from sugarcane harvesting in that, in the latter, the leaves are usually burned before harvesting. Since this is not possible for sweet sorghum, leaves and panicle are usually collected together with the stalk when using sugarcane harvesters. This can lead to contamination problems during the extraction of the juice, and does not permit separate processing of the grain (Lingle, 2010). Equipment for sweet sorghum harvesting and pressing have been proposed and tested at research institutions (Monroe et al., 1983; Nuese and Hunt, 1983; Monroe and Summer, 1985; Rains and Cundiff, 1990; Kundiyana et al., 2006; Pari, 2010). Agro-equipment companies John Deere and Case IH have also developed prototype sweet sorghum harvesters (Kennedy, 2011), but no commercial equipment is yet available.

5. IMPROVEMENT OF BIOMASS SORGHUM FOR BIOENERGY APPLICATIONS

Many of the traits that contribute to the overall yield potential of a crop are important irrespective of the end purpose of the product. For example, resistance to pests and diseases and tolerance to abiotic stresses such as drought and marginal lands are advantageous for any production system to provide yield stability and minimize the amount of such inputs as pesticides, fertilizer, and irrigation. There are some traits that are of specific interest for improvement for bioenergy production systems. Those include factors contributing to green biomass, sugar yield, and cell wall digestibility.

5.1. Green Biomass Yields

5.1.1. Plant Heigh
The height of the plant is determined by the length and number of internodes. Several genes that influence both internode length and duration of the growing period have been identified. The rate of node production is similar among genotypes (Sieglinger, 1936), therefore the number internodes that a plant will produce will be determined by the length of its vegetative growth stage. Sorghum originated in the tropics and evolved to have its reproductive cycle controlled by a combination of day length and temperature (photoperiod sensitive; Curtis, 1968; Doggett, 1988). Days of fewer than 12 hours trigger the switch from vegetative to reproductive stage in varieties that do not have mutations in the genes controlling this mechanism. The allelic state of the *Maturity* (*Ma*) genes determine whether a genotype is photoperiod sensitive or insensitive. In general, the dominant form of the genes causes lateness, but the different loci interact both among themselves and with temperature to produce a range of maturity responses (Quinby, 1974; Rooney and Aydin, 1999). Of the known *Ma* genes, Ma_5 and Ma_6 are in the process to be patented for use in increasing plant productivity and biomass accumulation (Mullet et al., 2010). The photoperiod-sensitivity trait is of interest for lignocellulosic biomass production, as plants will continue to grow and accumulate biomass during the entire growing season. For the production of sugars, however,

a balance between maximizing biomass production and allowing the plant to reach grain maturity needs to be achieved, as sugars in the stem juice reach their maximum concentration between flowering and physiological maturity of the grain.

Internode length is controlled by four genes, named *Dwarfing* (*Dw*) 1-4 (Quinby and Karper, 1954). The *Dw* genes have partial dominance for tallness, and their effects are additive in nature. Sorghum genotypes with all four *Dw* genes in their dominant form can reach over 4 m tall, while cultivars recessive for all four genes reach only 0.45–0.6 m in height. Of the four *Dw* genes, only *Dw3* has been cloned (Multani et al., 2003). Molecular markers for the wild type, *dw3*, and revertant alleles have been developed. The *Dw2* and *Dw3* genes have been mapped to sorghum chromosomes 6 and 7, respectively (Klein et al., 2008; Murray et al., 2008). The exact genomic locations of *Dw1* and *Dw4* have not yet been reported. The knowledge of the allelic status of parental lines could facilitate the production of biomass hybrid seed, as tall hybrids heterozygous for the *Dw* genes could be created from short parents with contrasting *Dw* genes. This would allow mechanized harvest of the hybrid seed, increasing the cost efficiency of the operation.

5.1.2. Sugar Concentration

Although early studies suggested a single dominant gene controlling the sweet stem trait (Ayyangar et al., 1936), it is now commonly accepted that the sugar accumulation in the stem is a quantitative trait controlled by multiple genes with additive effects. Progeny derived from both sweet × sweet and sweet × non-sweet crosses contain lines with higher sugar content than the parents (Natoli et al., 2002; Li, 2004; Murray et al., 2008; Ritter et al., 2008), indicating that further improvements in the sugar content are still possible. There is on-going research aimed at identifying the genes responsible for the control of sugar accumulation in the stem. Several QTL for sugar accumulation-related traits have been identified in segregant populations resulting from sweet x non-sweet crosses (Bian et al., 2006; Murray et al., 2008; Ritter et al., 2008). However, the QTLs only partially explain the observed variation of the traits, and only a small subset overlap among studies, indicative of the strong influence of environmental factors on the expression of the traits.

5.1.3. Cell Wall Composition

The lignocellulosic biomass of sorghum consists mostly of cell walls. The composition of the cell wall therefore influences the efficiency of the processing of the biomass for fuel production. Cell wall mutants with altered cell wall composition and structure may have advantageous processing traits. The best studied cell wall mutants of sorghum are the *brown midrib* (*bmr*) mutants (Porter et al., 1978; Xin et al., 2009). The mutations affect the amount and the composition of the lignin present in the cell wall. Enzymatic saccharification of the stover of some of the *bmr* mutants results in enhanced yields of fermentable sugars (Vermerris et al., 2007; Saballos et al., 2008). This can translate into increased production of ethanol and reduction of the costly pretreatment necessary for the biochemical processing of the biomass. However, some agronomic problems, such as lodging and yield penalties, have been reported as a result of the mutations (Pedersen et al., 2005; Oliver et al., 2005ab). The effect of the *bmr* mutation seems to be variable depending on the genetic background of the lines and hybrids in which the *bmr* trait is introduced, opening the possibility that careful selection of the lines in which the *bmr* trait is introduced can compensate for the fitness penalty associated with the mutations. At least four independent genes can confer the *bmr* phenotype (Bittinger et al., 1981; Saballos et al., 2008), but it is possible that additional *bmr* genes are present in the collections (Xin et al., 2009). The genes underlying the *bmr6* and the *bmr12* allelic groups have been identified. The mutants in the *bmr12* group present DNA lesions in the *caffeic acid O-methyl transferase* gene (Gene bank ID AY217766; Bout and Vermerris, 2003; Xin et al., 2008). The *bmr6* mutations are caused by DNA lesions in the *cinnamyl alcohol dehydrogenase* gene (Sb04g005950; Saballos et al., 2009; Sattler et al., 2009). The identification of the genes responsible for the *bmr6* and *bmr12* traits allowed the development of molecular markers that are now available to facilitate the introgression of the traits into bioenergy lines.

As mentioned above, the lignin polymer has a higher thermal value than cellulose. From a thermochemical processing standpoint, increased lignin content may be advantageous. The use of the naturally occurring variations in lignin content and transgenic manipulation to up-regulate genes in the lignin biosynthesis pathway are being investigated to improve the value of sorghum stover for co-firing applications (Sattler, *per. comm.*). However, there is some evidence that lignin concentration may be related to soot formation and changes in particle size of the processed biomass, which could reduce combustion efficiency (Bridgeman et al., 2007; Fitzpatrick et al., 2008). Further studies on the effects of increased lignin content in the processing characteristics of sorghum biomass are necessary before this strategy can be implemented.

Beside the lignin content, additional factors can have an effect on the processing characteristics of sorghum biomass. For example, the degree of crystallinity of the cellulose has been shown to impact the rate of enzymatic hydrolysis in other species (Yoshida et al., 2008; Hall et al., 2010). A high-throughput screen for monomeric sugar yield after hydrolysis has identified sorghum varieties with improved yield (Vandenbrink et al., 2010). Further studies on those lines can identify the factors that confer this yield advantage, so they can be used in the improvement of bioenergy lines.

6. CURRENT STATUS OF SORGHUM AS A BIOENERGY CROP

Within the spectrum of plants being considered as dedicated sources of bioenergy, sorghum is well positioned to become a major crop. Its high yield potential and tolerance to abiotic stress can allow it to produce economically attractive amounts of biomass with fewer inputs than most other grain or biomass crops. The versatility of the species allows it to fit well for all three bioenergy processing technologies, namely starch, sugar, and lignocellulosic biomass-based fuels. Because of this, the investment made in knowledge acquisition on the biology, agronomy, and genetic improvement of the species will be beneficial for all stages of the transition from starch-based to lignocellulosic bioenergy production.

The recognition of the potential of sorghum as a source of bioenergy has prompted investment in the generation of resources for its improvement. In the U.S., government support in the form of research grants allowed the complete sequencing of the sorghum genome (Paterson et al., 2009, available at: www.phytozome.net/sorghum) and the funding of numerous studies on the genetics, genomics, physiology, and agronomic requirements of the species. A germplasm collection with entries of wordwide origin representing the genetic diversity of the species is available from the Germplasm Resource Information Network

(GRIN, available at www.arsgrin.gov). Additionally, sorghum for food, feed, and forage applications has been utilized for more than a century in the U.S., and commercial seed companies are already familiar with the techniques of sorghum improvement and seed production.

Internationally, the tolerance of sorghum to harsh environments and its capacity to produce both food and energy have caught the interest of countries with growing populations and extensive areas of land with water limitations, especially in Asia and countries in the Mediterranean region. China and India are making strong investments in the development of sorghum (Li et al., 2004; Li and Chan-Halbrendt, 2009; Zhang et al., 2010, Reddy et al., 2007). ICRISAT in India has released bioenergy sweet sorghums varieties, and, in partnership with private enterprise, is working on the optimization of commercial sweet sorghum to ethanol production. Based on the number and variety of research articles published in Chinese academic journals, extensive research is being done in China on the development of varieties and hybrids, crop production, biomass processing, co-product production, and utilization, but little public information is available.

Some advances have been made in the commercial utilization of sorghum in bioenergy production. The National Sorghum Producers (NSP), through the Sorghum Checkoff, is working on the promotion of the use of sorghum grain in ethanol plants by educating growers and processing plants on the potential of sorghum for the ethanol industry. The NSP has also entered into a formal collaborative agreement with the Sweet Sorghum Ethanol Association (sseassociation.org) to expand the efforts to develop the sweet sorghum industry.

A commercial-scale plant with plans to use sweet sorghum as feedstock is being built in Ft. Lauderdale, FL, by Southeast Renewable Fuels. The plant is scheduled to be operational by the end of 2012, and it is expected that at least 10,000 ha will be involved in the production of feedstock for the plant (Kennedy, 2011). Abengoa Bioenergy has announced the restart of the Portales, NM, bioethanol plant, which plans to utilize sweet sorghum as its primary feedstock (abengoabioenergy.com, January 13, 2011). Biodimensions, in Memphis, TN, is installing a pilot-scale plant that will use sweet sorghum as its feedstock. Numerous commercial- and pilot-scale plants for bioethanol production with plans to use sorghum biomass as part of their feedstock mixture are operational or in the process of being built around the U.S. and Canada (ethanolproducer.com, January 27, 2011).

Lack of specialized biomass sorghum harvest equipment, limited availability of improved biomass sorghum hybrid seed, and the diversity of proposed harvest and processing technologies are still obstacles that need to be overcome to allow the adoption of sorghum as a biomass feedstock. Although sorghum has many advantageous traits and abundant resources for its improvement as a bioenergy crop, the development and expansion of this new use for the crop rests together with the fate of the bioenergy industry as a whole.

ACKNOWLEDGMENTS

Financial support from the Office of Science (BER), U.S. Department of Energy, grant DE-FG02-07ER64458, for research on improving sorghum biomass conversion is gratefully acknowledged.

REFERENCES

Almodares, A., Hadi, M.R. and Ahmadpour, H. (2009). Sorghum stem yield and soluble carbohydrates under different salinity levels. *Afr J Biotechnol, 7,* 4051-4055.

Almodares A, Sepahi A (1996). Comparison among sweet sorghum cultivars, lines and hybrids for sugar production. *Ann Plant Physiol, 10*, 50-55.

Almodares, A. and Hadi, M.R. (2009). Production of bioethanol from sweet sorghum: A review. *Afr J Agr Res, 4*, 772-780.

Amaducci, S., Monti, A. and Venturi, G. (2004). Non-structural carbohydrates and fibre components in sweet and fibre sorghum as affected by low and normal input techniques. *Ind Crops Prod, 20*, 111-118.

Antonopoulou, A., Ntaikou, I., Gavala, H.N., Skiadas, I.V., Angelopoulos K. and Lyberatos, G. (2007). Biohydrogen production from sweet sorghum biomass using mixed acidogenic cultures and pure cultures of *Ruminococcus albus*. *Global NEST J, 9,* 144–151.

Audilakshmi, S., Stenhouse, J.W., Reddy T.P. and. Prasad M.V. (1999). Grain mould resistance and associated characters of sorghum genotypes. *Euphytica, 107*, 91-103.

Audilakshmi, S., Mall, A.K., Swarnalathaa, M. and Seetharama N. (2010). Inheritance of sugar concentration in stalk (brix), sucrose content, stalk and juice yield in sorghum. *Biomass Bioenergy, 34*, 813-820.

Awika, J.M. and Rooney. L.W. (2004). Sorghum phytochemicals and their potential impact on human health. *Phytochem, 65*,1199-1221.

Ayyangar, G., Ayyar, M., Rao, V., and Nambiar, A. (1936). Mendelian segregation for juiciness and sweetness in sorghum stalk. *Madras Agric. J. 24,* 247.

Basu, P., Buttler, J. Leon, M. (2011). Biomass co-firing options on the emission reduction and electricity generation costs in coal-fired power plants. *Renew Energ, 36*, 282-288.

Bele, P. (2007). Economics of on-farm ethanol production using sweet sorghum. *M.S. Thesis.* The Graduate College, Oklahoma State University. Stillwater, OK.

Bellmer, D.D., Huhnke, R.L. and Kundiyana, D. (2008). Issues with In-Field Fermentation of Sweet Sorghum Juice. 2008 ASABE Annual International Meeting, Jun 29-Jul 2. Rhode Island Convention Center, Providence, RI.

Bian, Y-L., Seiji, Y., Miako, I., Cai, H-W. (2006). QTLs for sugar content of stalk in sweet sorghum (Sorghum bicolor L. Moech). *Agric Sci China, 5*, 736-744.

Bio-Energy Solutions, Inc. (2009). Grain sorghum in ethanol [PDF document]. http://www.sorghumcheckoff.com/userfiles/USCP%20Study%20_2_.pdf. Accessed Jan 03, 2011.

Bittinger, T.S., Cantrell R. P. and Axtell, J. D. (1981). Allelism tests of the *brownmidrib* mutants of sorghum. *J. Hered, 72*, 147–148.

Blade Energy Crops® (2010). Managing high biomass sorghum [PDF .document] www.bladeenergy.com/Bladepdf/Blade_SorghumMgmtGuide2010.pdf. Accessed Feb 11, 2011.

Blitzer, M. (1997). *Production of Sweet Sorghum for Syrup in Kentucky*. University of Kentucky Cooperative Extension Service [On-line publication Agr-122]. http://www.ca.uky.edu/agc/pubs/agr/agr122/agr122.htm.

Botha, T. and von Blottnitz, H. (2006). A comparison of the environmental benefits of bagasse-derived electricity and fuel ethanol on a life-cycle basis. *Energ Policy, 34*, 2654-2661.

Bout, S. and Vermerris, W. (2003). A candidate-gene approach to clone the sorghum *Brown midrib* gene encoding *caffeic acid O-methyltransferase*. *Mol Genet Genomics, 269*, 205–214.

Bridgeman, T.G., Darvell, L.I., Jones, J.M., Williams, P.T., Fahmi, R., et al., (2007). Influence of particle size on the analytical and chemical properties of two energy crops. *Fuel, 86,* 60–72

Bryan, W.L., Monroe, G.E. and Caussanel, P.M. (1985). Solid-phase fermentation and juice expression systems for sweet sorghum. *Trans ASAE (Am Soc Agric Eng), 28,* 268-274.

Carpita, N. and Gibeaut, D., (1993). Structural models of the primary cell walls in flowering plants: consistency of molecular structure with the physical properties of the walls during growth. *Plant J, 3,* 1-30.

Carpita, N. and McCann, M. (2000). The cell wall. In B. Buchanan, W. Gruissem and R. Jones (Eds.), *Biochemistry and Molecular Biology of Plants* (2nd edition, pp. 52-109). Somerset, NJ: John Wiley.

Chang, V. and Holtzapple, M. (2000). Fundamental factors affecting biomass enzymatic reactivity. *Appl Biochem Biotechnol, 84,* 5–38.

Channappagoudar, B.B., Biradar, N.R., Patil, J.B. and Hiremath, S.M. (2007). Assessment of sweet sorghum genotypes for cane yield, juice characters and sugar levels. *Karnataka J Agric Sci, 20,* 294-296.

Chernoglazov, V., Ermolova, O., Klyosov, A. (1988). Adsorption of high-purity endo-1,4-beta-glucanases from *Trichoderma reesei* on components of lignocellulosic materials: Cellulose, lignin, and xylan. *Enzyme Microb Technol, 10,* 503–507.

Chohnan, S., Nakane, M., Rahman, M.H., Nitta, Y., Yoshiura, T., Ohta H. and Kurusu Y. (2011). Fuel ethanol production from sweet sorghum using repeated-batch fermentation. *J Biosci Bioeng, in press*, doi:10.1016/j.jbiosc.2010.12.014.

Claassen, P.A.M., de Vrije, T. and Budde M.A. (2004). Biological hydrogen production from sweet sorghum by thermophilic bacteria. 2nd World Conference on Biomass for Energy, Industry and Climate Protection, May 10-14. Rome, Italy.

Converse, A. (1993). Substrate factors limiting enzymatic hydrolysis. In: J. Saddler (ed.), *Bioconversion of Forest and Agricultural Plant residues* (pp. 93-106). Wallingford, UK: CAB.

Delmer, D. (1999). Cellulose biosynthesis: Exciting times for a difficult field of study. *Ann Rev Plant Physiol Plant Mol Biol, 50,* 245-276.

Dias, M.O., , Modesto, M., Ensinas, A.V., Nebra, S.A., Filho R.M. and Rossell C.E. (2010). Improving bioethanol production from sugarcane: evaluation of distillation, thermal integration and cogeneration systems. *Energy, in press*, doi:10.1016/j.energy.2010.09.024.

Doggett, H. (1988). *Sorghum* (2nd ed.). New York, NY: John Wiley.

Dolciotti, I., Mambelli, S., Grandi, S., and Venturi, G. (1998). Comparison of two sorghum genotypes for sugar and fiber production. *Industrial Crops Products*, 7, 265–272.

Eckhoff, S.R., Bender, D.A., Okos, M.R. and Peart R.M. (1985). Preservation of chopped sweet sorghum using sulfur dioxide. *Trans ASAE (Am Soc Agric Eng), 28,* 606-609.

Eickhout, B., Bouwman, A.F., van Zeijts, H. (2006). The role of nitrogen in world food production and environmental sustainablility. *Agricult Ecosys Environ, 116*, 4-14.

Eiland, B.R,. Clayton J.E. and Bryan, W.L. (1983). Losses of fermentable sugars in sweet sorghum during storage. *Trans ASAE (Am Soc Agric Eng), 26*, 1596–1600.

Esela, J.P. Frederiksen, R.A. and Miller, F.R. (1993). The association of genes controlling caryopsis traits with grain mold resistance in sorghum. *Phytopathology, 83*, 490-495.

Farrell, A.E., Plevin, R.J., Turner, B.T., Jones, A.D., O'Hare, M., et al., (2006). Ethanol can contribute to energy and environmental goals. *Science, 311*, 506-508.

Fitzpatrick, E.M., Jones, J.M., Pourkashanian, M., Ross, A.B., Williams, A., et al., (2008). Mechanistic aspects of soot formation from the combustion of pine wood. *Energy Fuels, 22*, 3771–3778.

Gao, C., Zhai, Y., Ding, Y. and Wu, Q. (2009). Application of sweet sorghum for biodiesel production by heterotrophic microalga *Chlorella protothecoides*. *Appl Energ, 3*, 756-761.

Gardner K., Blackwell J. (1974). The structure of native cellulose. *Biopolymers, 13,* 1975-2001.

Gibbons, W.R., Westby C.A. and Dobbs T.L. (1986). Intermediate-scale, semicontinuous solid-phase fermentation process for production of fuel ethanol from sweet sorghum. *Appl Environ Microbiol, 51*, 115-122.

Habyarimana, E., Laureti, D., De Ninno, M. and Lorenzoni C. (2004a). Performances of biomass sorghum [*Sorghum bicolor* (L.) Moench] under different water regimes in Mediterranean region. *Ind Crop Prod, 20*, 23-28.

Habyarimana E., Bonardi, P., Laureti, D., Di Bari, V., Cosentino, S. and Lorenzoni. C. (2004b). Multilocational evaluation of biomass sorghum hybrids under two stand densities and variable water supply in Italy. *Ind Crop Prod, 20*, 3–9.

Hall, M., Bansal, P., Lee, J.H., Realff, M.J. and Bommarius, A.S. (2010). Cellulose crystallinity – a key predictor of the enzymatic hydrolysis rate. *FEBS J, 277*, 1571-1582.

Hill, J., Nelson, E., Tilman, D., Polasky, S. and Tiffany, D. (2006). Environmental, economic, and energetic costs and benefits of biodiesel and ethanol biofuels. *Proc Nat Acad Sci. U.S.A., 103*,11206-11210

Houtman, C., Atalla, R. (1995). Cellulose-lignin interactions. *Plant Physiol, 107*, 977-984.

James, M.G., Denyer, K. and Myers, A.M. (2003). Starch synthesis in the cereal endosperm. *Curr Opin Plant Biol, 6*, 215-222.

Keating, B. A., Webster, A. J., Hoare, C. P., and Sutherland, R. F. (2004). Observations of the harvesting, transporting and trial crushing of sweet sorghum in a sugar mill. 2004 Conference of the Australian Society of Sugar Cane Technologists, May 4-7. Brisbane, Queensland, Australia.

Kennedy, L. (2011). Sorghum's sweet future. *Sorghum Grower, 5,* 8-11.

Kim. S. and Dale, B.E. (2008). Effects of nitrogen fertilizer application on greenhouse gas emissions and economics of corn production. *Environ Sci Technol, 42,* 6028-6033.

Kim M. and Day D.F. (2010). Composition of multiple feedstocks suitable for extended ethanol production at Louisiana sugar mills. 32nd Symposium on Biotechnology for Fuels and Chemicals, April 19-22. Clearwater Beach, FL.

Klein, R.R., Mullet, J.E., Jordan, D.R., Miller, F.R., Rooney, W.L., et al., (2008). The effect of tropical sorghum conversion and inbred development on genome diversity as revealed by high-resolution genotyping. *Crop Sci, 48,* S12–S26.

Kundiyana, D., Bellmer, D., Huhnke, R., and Wilkins, M. (2006). "Sorganol®": Production of Ethanol from Sweet Sorghum. 2006 American Society of Agricultural and Biological Engineers Annual International Meeting, July 9-12. Portland Convention Center, Portland, Oregon.

Laidlaw, H.K.C., Mace, E.S., Williams, S.B., Sakrewski, K., Mudge, A.M., et al., (2010). Allelic variation of the β-, γ- and δ-kafirin genes in diverse *Sorghum* genotypes. *Theor Appl Genet,121*, 1227-1237.

Laser, M., Larson, E., Dale, B., Wang, M., Greene, N., et al., (2009). Comparative analysis of efficiency, environmental impact, and process economics for mature biomass refining scenarios. *Biofuels Bioprod Biorefin, 3*, 247-270.

Leslie, J.F., Zeller, K.A., Lamprecht, S.C., Rheeder, J.P. and Marasas, W.F. (2005). Toxicity, pathogenicity, and genetic differentiation of five species of Fusarium from sorghum and millet. *Phytopathology, 95*, 275–83.

Li, G., Gu, W., and Chapman, K., eds. (2004). *Sweet Sorghum*. Beijing, China. C and C Joint Printing Co.

Li, S-Z. and Chan-Halbrendt, C. (2009). Ethanol production in (the) People's Republic of China: Potential and technologies. *Appl Energ, 86*, S162-169.

Lingle, S. (2010). Opportunities and Challenges of Sweet Sorghum as a Feedstock for Biofuel. In Eggleston, G. (ed) *Sustainability of the Sugar and Sugar–Ethanol Industries* (pp. 177–188). Washington, DC. ACS Symposium Series; American Chemical Society. [on-line edited book]. doi: 10.1021/bk-2010-1058.ch011.

Lipinsky, E.S. and Kresovich, S. (1980). Sorghums as energy crops. In The Bio-Energy Council (Ed.) *Proceedings of the Bio-Energy '80 World Congress and Exposition. April 21-24, 1980. Atlanta, GA.* (pp. 91-93). Washington, DC.

Lu, Y. and Mosier, N. (2008). Current technologies for fuel ethanol production from lignocellulosic plant biomass. In W. Vermerris (Ed.), *Genetic improvement of bioenergy crops* (pp. 161-182). New York, NY: Springer.

Mace, E.S., Rami, J-F., Bouchet, S., Klein, P.E., Klein, R.R., et al., (2009). A consensus genetic map of sorghum that integrates multiple component maps and high-throughput Diversity Array Technology (DArT) markers. BMC Plant Biol [On-line serial], 9. doi:10.1186/1471-2229-9-13.

Mathews, K.H. and McConnell, M.J. (2009). Ethanol co-product use in U.S. cattle feeding. [United States Department of Agriculture, Report FDS-09D-01]. http://ae2030.hautetfort.com/media/00/02/1628713052.pdf

Mask, P.L. and Morris, W.C. (1991). *Sweet Sorghum Culture and Syrup Production*. Alabama Cooperative Extension Service, Auburn University [on-line Circular ANR-625]. http://www.aces.edu/pubs/docs/A/ANR-0625/.

Maskina, M.S., Power, J.F., Doran, J.W. and Wilhelm, W.W. (1993). Residual effects of no-till crop residues on corn yield and nitrogen uptake. *Soil Sci Soc Am J, 57*, 1555-1560.

McBee, G.G., Creelman, R.A. and Miller F.R. (1988). Ethanol yield and energy potential of stems from a spectrum of sorghum biomass types. *Biomass, 17,* 203-211.

Mei, X., Liu R. and Shen F. (2008). Physiological property of postharvest sweet sorghum stalk and its modified atmosphere packaging storage. *Trans Chinese Soc Agric Eng, 07*.doi: CNKI:SUN:NYGU.0.2008-07-036.

Miller, F.R., Quinby, J.R. and Cruzdao, H.J. (1968). Expression of known maturity genes of sorghum in temperate and tropical environments. *Crop Sci, 8*, 675-677

Monroe, G.E., Summer H.R. and Hellwig, R.E. (1983). A Leaf-Removal Principle for Sweet Sorghum. *Trans ASAE (Am Soc Agric Eng), 26,* 1006-1010.

Monroe, G.E. and Summer H.R. (1985). Preservation of chopped sweet sorghum using sulfur dioxide. *Trans ASAE (Am Soc Agric Eng), 28,* 562-567.

Mooney, C., Mansfield, S., Touhy, M. and Saddler, J. (1998). The effect of initial pore volume and lignin content on the enzymatic hydrolysis of softwoods. *Biores Technol, 94,* 113–119.

Mosier, C., Wyman, B., Dale, R., Elander, Y., Lee, M., Holtzapple and Ladisch, M. (2005). Features of promising technologies for pretreatment of lignocellulosic biomass. *Biores Technol, 96,* 673–686.

Mullet, J.E., Rooney, W.L., Klein, P.E., Morishige, D., Murphy, R., et al., (2010). Discovery and utilization of sorghum genes (MA5/MA6). United States patent application, Publication number: US 2010/0024065 A1.

Multani, D.S., Briggs, S.P., Chamberlain, M.A., Blakeslee, J.J., Murphy, A.S., et al., (2003). Loss of an MDR transporter in compact stalks of maize *br2* and sorghum *dw3* mutants. *Science, 302,* 81-84.

Murray, S.C., Sharma, A., Rooney, W.L., Klein, P.E., Mullet, J.E., et al., (2008). Genetic improvement of sorghum as a biofuel feedstock I: quantitative loci for stem sugar and grain nonstructural carbohydrates. *Crop Sci, 48,* 2165-2179.

Nathan, R. A., ed. (1978). *Fuels from sugar crops : systems study for sugarcane, sweet sorghum, and sugar beets.* Honolulu, HI. University Press of the Pacific.

Natoli, A., Gorni, C., Chegdani, F., Ajmone Marsan, P., Colombi, C., et al., (2002). Identification of QTLs associated with sweet sorghum quality. *Maydica, 17,* 311-322.

Newman, Y., Erickson, J., Vermerris, W., and Wright D. (2010). *Forage Sorghum (Sorghum bicolor): Overview and Management.* University of Florida IFAS Extension [On-line publication SS-AGR-333]. http://edis.ifas.ufl.edu/ag343.

Nichols, N.N and Bothast, R.J. (2008). Production of ethanol from grain. In W. Vermerris (Ed.), *Genetic improvement of bioenergy crops* (pp. 75-88). New York, NY: Springer.

Nuese, G.A. and Hunt, D.R. (1983). Field Harvest of Sweet Sorghum Juice. *Trans ASAE (Am Soc Agric Eng), 26,* 656-660.

Oliver, A., Pedersen, J., Grant, R. and Klopfenstein, T. (2005a). Comparative effect of the sorghum *bmr-6* and *bmr-12* genes I. Forage sorghum yield and quality. *Crop Sci, 45,* 2234–2239.

Oliver, A., Pedersen, J., Grant, R., Klopfenstein, T. and Jose, D. (2005b). Comparative effect of the sorghum *bmr-6* and *bmr-12* genes II. Grain yield, stover yield and stover quality in grain sorghum. *Crop Sci, 45,* 2240–2245.

Palonen, H., Tjerneld, F., Zacchi, G. and Tenkanen, M. (2004). Adsorption of *Trichoderma reesei* CBH I and EG II and their catalytic domains on steam pretreated soft wood and isolated lignin. *J Biotechnol, 107,* 65–72.

Paterson, A.H., Bowers, J.E., Bruggmann, R., Dubchak, I., Grimwood, J., et al., (2009). The *Sorghum bicolor* genome and the diversification of grasses. *Nature, 457,* 551-556.

Pari, L., Grassi. G., Sénéchal, S., Capaccioli, S. and Cocchi, M. (2008). State of the art: Harvest, storage and logistics of the sweet sorghum. 16th European Biomass Conference and Exhibition, Jun 2-6. Valencia, Spain.

Patil, S.L. (2007). Performance of sorghum varieties and hybrids during postrainy season under drought situations in Vertisols in Bellary, India. J SAT Agric Res [On-line serial], 5(1).

Pedersen, J. F., Vogel K. P. and Funnell, D. L. (2005). Impact of reduced lignin on plant fitness. *Crop Sci, 45,* 812–819.

Pfeiffer, T.W., Bitzer, M.J., Toy, J.J. and Pedersen, J.F. (2010). Heterosis in sweet sorghum and selection of a new sweet sorghum hybrid for use in syrup production in appalachia. *Crop Sci, 50,* 1788-1794.

Porter, K. S., Axtell, J. D., Lechtenberg, V. L., and Colenbrander, V. F. (1978). Phenotype, fiber composition, and in vitro dry matter disappearance of chemically induced *brown midrib* (*bmr*) mutants of sorghum. *Crop Sci. 18,* 205–208.

Prihar, S.S and Stewart, B.A (1991). Sorghum harvest index in relation to plant size, environment and cultivar. *Agron J, 83,* 603-608.

Prom, L.K. and Erpelding, J. (2009). New sources of grain mold resistance among sorghum accessions from Sudan. *Trop Subtrop Agroecosyst, 10,* 457-463.

Quaranta, F., Belocchi, A., Bentivenga, G., Mazzon, V. and Melloni, S. (2010). Fibre sorghum: influence of harvesting period and biological cycle on yield and dry matter in some hybrids. *Maydica, 55,* 173-177

Quinby, J.R. (1974). *Sorghum improvement and the genetics of growth.* College Station, TX. Texas AandM University Press.

Quinby, J.R. and and Karper, R.E. (1954). Inheritance of height in sorghum. *Agron J, 46,* 211-216.

Quintero, J.A., Montoya, M.I., Sanchez, O.J., Giraldo, O.H. and Cardona, C.A. (2008). Fuel ethanol production from sugarcane and corn: Comparative analysis for a Colombian case. *Energy, 33,* 385-399.

Ragauskas, A.J., William, C.K., Davison, B.H., Britovsek, G., Cairney, J., et al., (2006). The path forward for biofuels and biomaterials . *Science, 311,* 484-489.

Rains, C.G. and Cundiff, J.S. (1993). Field Harvest of Sweet Sorghum Juice. *App Eng Agric, 9,* 15-20.

Rajendran, C., Ramamoorthy, K. and Backiyarani, S. (2000). Effect of deheading onjuice quality characteristics and sugar yield of sweet sorghum. *J Agron Crop Sci, 85,* 23-26.

Ramos, L. (2003). The chemistry involved in the steam treatment of lignocellulosic materials. *Quim Nova, 26,* 863-871.

Ramu, P., Kassahun, B., Senthilvel, S., Kumar, A.C., Jayashree, B., et al., (2009). Exploiting rice-sorghum synteny for targeted development of EST-SSRs to enrich the sorghum genetic linkage map. *Theor Appl Genet, 119,* 1193-204.

Raveendran K. and Ganesh, A. (1996). Heating value of biomass and biomass pyrolysis products. *Fuel, 75,* 1715-1720.

Reddy, B., Ramesh, S., Reddy, P.S., Ashok Kumar, A.A., Sharma, K.K., Karuppan Chetty, S.M., and Palaniswamy, A.R. (2007). *Sweet Sorghum: Food, Feed, Fodder and Fuel Crop.* Patancheru , Andhra Pradesh, India. International Crops Research Institute for the Semi-Arid Tropics.

Ritter, K., Jordan, D., Chapman, S., Godwin, I., Mace, E. and McIntyre, L. (2008). Identification of QTL for sugar-related traits in a sweet x grain sorghum (*Sorghum bicolor* L. Moech) recombinant inbred population. *Mol. Breed, 22,* 367-384.

Rocateli A.C., Raper R. L., Olander B., Pruitt M., Schwab E. B. (2009). Rapidly Drying Sorghum Biomass for Potential Production. International Commission of Agricultural and Biological Engineers, Section V. Conference *"Technology and Management to Increase the Efficiency in Sustainable Agricultural Systems"*, Sept 1-4. Rosario, Argentina,

Rodriguez-Herrera, R., Waniska, R. D. and Rooney, W. (1999). Antifungal proteins and grain mold resistance in sorghum with nonpigmented testa. *J Agric Food Chem, 47,* 4802-4806.

Rooney, L. W. and Pflugfelder, R. L. (1986). Factors Affecting Starch Digestibility with Special Emphasis on Sorghum and Corn. *J. Anim Sci, 63,* 1607-1623.

Rooney, W. L. and Aydin, S. (1999). Genetic control of a photoperiod-sensitive response in *Sorghum bicolor* (L.) Moench. *Crop Sci, 39,* 397–400.

Rooney, W., Blumenthal, J., Bean, B., and Mullet, J. (2007). Designing sorghum as a dedicated bioenergy feedstock. *Biofuels Bioprod Bioref, 1,* 147–157.

Saballos, A. (2008). Development and utilization of sorghum as a bioenergy crop. In W. Vermerris (Ed.), *Genetic improvement of bioenergy crops* (pp. 211-248). New York, NY: Springer.

Saballos, A., Vermerris, W., Rivera, L. and Ejeta, G. (2008). Allelic association, chemical characterization and saccharification properties of *brown midrib* mutants of sorghum (*Sorghum bicolor* (L.) Moench). *BioEnerg. Res, 1,* 193-204.

Saballos, A., Ejeta, G., Sanchez, E., Kang, C. and Vermerris, W. (2009). A genome-wide analysis of the cinnamyl alcohol dehydrogenase family in sorghum (*Sorghum bicolor* (L.) Moench) identifies *SbCAD2* as the *Brown midrib6* gene. *Genetics, 181,* 783-795.

Sakellariou-Makrantonak, M., Papalexis D., Nakos, N. and Kalavrouziotis, I.K. (2007). Effect of modern irrigation methods on growth and energy production of sweet sorghum (var. Keller) on a dry year in Central Greece. *Agr Water Manage, 90,* 181-189.

Sang, Y., Bean, S., Seib, P.A., Pedersen, J. and Shi, Y-C. (2008). Structure and Functional Properties of Sorghum Starches Differing in Amylose Content. *J Agric Food Chem, 56,* 6680–6685.

Sattler, S., Saathoff, A.J., Palmer, N.A., Funnell-Harris, D.L., Sarath, G., et al., (2009). A nonsense mutation in a cinnamyl alcohol dehydrogenase gene is responsible for the sorghum *brown midrib 6* phenotype. *Plant Physiol, 150,* 584-595.

Sattler S.E., Singh, J., Haas, E.J., Guo, L., Sarath G., et al., (2009). Two distinct *waxy* alleles impact the granule-bound starch synthase in sorghum. *Mol Breed, 24,* 349-359.

Schmer MR, Vogel KP, Mitchell RB, Perrin RK (2008). Net energy of cellulosic ethanol from switchgrass. *Proc Natl Acad Sci U S A, 105,* 464-469.

Schmidt, J., Sipocz, J., Kaszás, I., Szakács, G., Gyepes A., and Tengerdy R.P. (1997). Preservation of sugar content in ensiled sweet sorghum. *Biores Biotechnol, 60,* 9-13.

Schwietzke, S., Ladisch, M., Russo. L., Kwant, K., Mäkinen, T., et al., (2008). Analysis and identification of gaps in research for the production of second-generation liquid transportation biofuels [On-line report IEA Bioenergy:T41(2): 2008:01]. http://www.ftconferences.com/userfiles/file/Berndes%20Goran%20Gaps_in_the_Researc h_of_2nd_Generation_Transportation_Biofuels.pdf

Sherwood, S. P. (1923). Starch in sorghum juice. *Ind Eng Chem, 15,* 727–728.

Sieglinger, J. B. (1936). Leaf number of sorghum stalks. *J Amer Soc Agron, 28,* 636.

Smith, C. W., and Frederiksen, R. A., eds. (2000). *Sorghum : origin, history, technology, and production*. New York, NY: John Wiley.

Stephens, J.C. (1924). A second factor for subcoat in sorghum seed. *J American Soci Agron, 25,* 340-342.

Teetor, V.H., Duclos, D.V., Wittenberg, E.T., Young, K.M., Chawhuaymak, J., et al., (2010). Effects of planting date on sugar and ethanol yield of sweet sorghum grown in Arizona. *Ind Crop Prod, in press.* doi:10.1016/j.indcrop.2010.09.010.

Tenenbaum, D.J. (2008). Food vs. fuel: Diversion of crops could cause more hunger. *Environ Health Perspect, 116,* A254-257.

Tew, T., Cobill, R. and Richard, E. (2008). Evaluation of sweet sorghum and sorghum x sudangrass hybrids as feedstocks for ethanol production. *Bioenerg Res, 1,* 147-152.

Tsuchihashi, N. and Goto,Y. (2004). Cultivation of sweet sorghum (*Sorghum bicolor* (l.) moench) and determination of its harvest time to make use as the raw material for fermentation, practiced during rainy season in dry land of Indonesia. *Plant Prod Sci, 7,* 442-448.

Vandenbrink, J.P., Delgado, M.P., Frederick J.R. and Feltus F.A. (2010). A sorghum diversity panel biofuel feedstock screen for genotypes with high hydrolysis yield potential. *Ind Crop Prod,* 31, 444-448.

Venuto, B. and Kindiger, B. (2008). Forage and biomass feedstock production from hybrid forage sorghum and sorghum–sudangrass hybrids. *Grassland Sci, 54,* 189-196.

Vermerris, W., Saballos, A., Ejeta, G., Mosier, N.S., Ladisch, M.R. and Carpita, N.C. (2007). Molecular breeding to enhance ethanol production from corn and sorghum stover. *Crop Sci, 47,* S145-153.

Wilhelm, W.W., Johnson, J.M., Hatfield, J.L., Voorhees, W.B. and Linden, D.R. (2004). Crop and soil productivity response to corn residue removal: a literature review. *Agron J, 96,* 1-17.

Wong, J.H., Lau, T., Cai, N., Singh, J., Pedersen, J.F., et al., (2009). Digestibility of protein and starch from sorghum (*Sorghum bicolor*) is linked to biochemical and structural features of grain endosperm. *J Cereal Sci. 49,* 73-82.

Wong, J.H., Marx, D.B., Wilson, J.D., Buchanan, B.., Lemaux. P., et al., (2010). Principal component analysis and biochemical characterization of protein and starch reveal primary targets for improving sorghum grain. *Plant Sci, 179,* 598-611.

Wortmann, C.S., Liska, A.J., Ferguson, R.B., Lyon, D.J., Klein, R.N., et al., (2010). Dryland performance of sweet sorghum and grain crops for biofuel in Nebraska. *Agron J, 102,* 319-326.

Wu, X., Zhao, R., Bean, S.R., Seib, P.A., McLaren, J.S., et al., (2007). Factors impacting ethanol production from grain sorghum in the dry-grind process. *Cereal Chem, 84,* 130-136.

Wu, F. and Munkvold, G.P. (2008). Mycotoxins in ethanol co-products: modeling economic impacts on the livestock industry and management strategies. *J Agric Food Chem, 56,* 3900–3911.

Wu, X., Staggenborg, S., Propheter, J.L., Rooney, W.L., Yu J. and Wang D.(2010). Features of sweet sorghum juice and their performance in ethanol fermentation. *Ind Crop Prod, 31,* 164-170.

Xiang W. (2009). Identification of two interacting quantitative trait loci controlling for condensed tannin in sorghum grain and grain quality analysis of a sorghum diverse collection. *M.S. Thesis.* Department of Agronomy, College of Agriculture. Kansas State University. Manhattan, KS.

Xin. Z., Wang. M.L., Barkley. N.A., Burow. G., Franks. C., et al., (2008). Applying genotyping (TILLING) and phenotyping analyses to elucidate gene function in a chemically induced sorghum mutant population. *BMC Plant Biol, 8,* 108-140.

Xin, Z., Wang, M.L., Burow, G., Burke, J. (2009). An induced sorghum mutant population suitable for bioenergy research. *BioEnerg Res, 2,* 10-16

Xiong, S., Zhang, Q., Zhang, D., Olsson, R. (2008). Influence of harvest time on fuel characteristics of five potential energy crops in northern China. *Bioresour. Technol. 99,* 479–485.

Yonemaru, J-I., Ando, T., Mizubayashi, T., Kasuga, S., Matsumoto, T., et al., (2009). Development of genome-wide sequence repeat markers using whole-genome shotgun sequences of sorghum (*sorghum bicolor* (L.) Moench). *DNA Res, 16,* 187-93.

Yoshida, M., Liu, Y., Uchida, S., Kawarada, K., Ukagami, Y., et al., (2008). Effects of cellulose crystallinity, hemicellulose, and lignin on the enzymatic hydrolysis of *Miscanthus sinensis* to monosaccharides. *Biosci Biotech Biochem, 72,* 805-810.

Zhan, X., Wang, D., Tuinstra, M. R., Bean, S., Seib, P. A., et al., (2003). Ethanol and lactic acid production as affected by sorghum genotype and location. *Ind Crop Prod, 18,* 245-255.

Zhang, C., Xie, G., Li, S., Ge, l. and He, T. (2010). The productive potentials of sweet sorghum ethanol in China. *Appl Energ, 87,* 2360-2368.

Zhao, R., Bean, S.R., Wang, D., Park, S.H., Schober, T.J., et al., (2009). Small-scale mashing procedure for predicting ethanol yield of sorghum grain. *J Cereal Sci, 49,* 230-238.

INDEX

A

access, 21, 22
accessibility, 36, 145, 148, 151
accessions, 5, 167
acetic acid, 25, 57, 69
acid, ix, 11, 14, 15, 23, 25, 39, 52, 54, 56, 57, 68, 71, 72, 73, 75, 79, 80, 109, 110, 111, 112, 114, 115, 123, 124, 125, 126, 127, 130, 133, 136, 137, 139, 150, 160, 163
acidic, 110, 127
acidity, 46, 126
activated carbon, 24
adaptability, 46
adaptation, ix, 109, 111, 115, 142
additives, 15
adjustment, 71, 116, 117
adsorption, 58
advancements, 25
adverse weather, 152
aflatoxin, 147
Africa, viii, ix, 2, 3, 4, 5, 6, 8, 12, 13, 14, 15, 24, 33, 39, 40, 42, 65, 110, 129, 130, 141, 142, 153
agar, 61, 132, 133
air quality, 50
air temperature, 47
Alaska, 125
Albania, 5
albumin, 11
alcohol production, 17, 18, 50, 60
alcohols, 16, 148
aldehydes, 16, 17, 52
algae, 56, 148
Algeria, 8
algorithm, 126
alkalinity, 47, 66
allele, 146

allergy, 11
alternative energy, 18
amino, 11, 25, 35
amino acid, 11, 25, 35
amino acids, 35
ammonia, 68, 69, 71, 76, 79, 80
ammonium, 59, 67
amylase, 14, 15, 19, 20, 21, 25, 27, 29, 31, 33, 34, 37, 39, 52, 54, 63
anaerobe, 74
anaerobic digestion, 78
animal feed, vii, 1, 3, 5, 8, 12, 13, 16, 17, 19, 20, 23, 31, 43, 49, 50, 141, 144, 145, 151
Animal Food, vii, 1
antibiotic, vii, x, 129, 130, 131, 132, 133, 139
antibiotic resistance, vii, x, 129, 130, 131, 139
antioxidant, x, 16, 129, 130, 131, 132, 134, 137, 138, 139
apex, 126
aqueous solutions, 126
Argentina, 3, 5, 6, 8, 9, 168
arid regions, vii, 1, 3, 142
aromatics, 134
arrests, 142
ascorbic acid, 37, 131, 134
Asia, ix, 2, 3, 5, 8, 12, 13, 14, 16, 65, 110, 129, 141, 161
atmosphere, 75, 76, 165
atmospheric pressure, 79
avoidance, 15

B

Bacillus subtilis, x, 19, 129, 131, 136
bacteria, x, 51, 57, 75, 81, 129, 135, 151, 163
bacterial strains, 133, 137
bacterium, 74, 80, 81

Bangladesh, 109, 110
barley, vii, 1, 2, 15, 24, 29, 33, 34, 61, 112
barriers, 76
base, ix, 146
beef, 16
beer, 15, 24, 25, 34, 38, 39, 41, 42, 143
Beijing, 165
beneficial effect, 73
benefits, ix, 16, 60, 69, 163, 164
beverages, vii, 15
biochemical processes, 151
biochemistry, 51, 131
bioconversion, 20, 58, 79, 80
biodiesel, 62, 72, 73, 78, 80, 148, 164
biodiversity, 66
bioenergy, viii, x, 65, 66, 78, 141, 142, 145, 147, 148, 155, 156, 158, 160, 161, 165, 166, 168, 169, 170
bioethanol, vii, ix, 1, 3, 12, 18, 20, 24, 29, 31, 38, 41, 61, 62, 63, 64, 77, 78, 79, 80, 144, 161, 162, 163
biofuel, viii, 46, 50, 63, 65, 68, 76, 77, 78, 79, 143, 147, 166, 169
bio-fuel, viii
bio-fuel, 45
biological activities, vii, x, 129, 136
biomass, x, 46, 49, 50, 52, 56, 57, 58, 61, 68, 69, 71, 73, 74, 75, 76, 78, 79, 80, 81, 141, 142, 145, 147, 148, 149, 150, 151, 152, 154, 155, 156, 157, 158, 159, 160, 161, 162, 163, 164, 165, 166, 167, 169
biomass growth, 56
biomaterials, 167
biopolymer, 25
biopolymers, 57, 80
biosynthesis, 125, 160, 163
biotechnology, 63
birds, 17, 37
birefringence, 54
bleaching, 23
blowing agent, 20
bonding, 37
bonds, 25, 146
Botswana, 5
Brazil, 3, 5, 6, 9, 110, 142
breakdown, 23, 37
breeding, ix, 4, 20, 24, 111, 126, 144, 146, 147, 148, 153, 169
brewing industries, vii, 1, 3
Brno, 131
Burkina Faso, 5, 9, 13, 40
Burundi, 3, 5
butyl ether, viii, 45, 49, 61, 62
by-products, 52

C

Ca^{2+}, 111, 121, 125, 126
calcium, 126
calorie, 13
Cambodia, 110
Cameroon, 5
cancer, 73
candidates, 142, 155
carbohydrate, viii, 39, 45, 47, 49, 50, 61, 62, 68, 77, 154
carbohydrates, 10, 49, 58, 61, 62, 63, 68, 79, 81, 162, 166
carbon, 20, 50, 52, 56, 59, 63, 78, 151
carbon dioxide, 20, 52
carbon emissions, 50
carbon monoxide, 151
cardiovascular disease, 16, 73
carotene, 16
carotenoids, 130
case studies, 126
case study, 78
casting, 42
cation, 131, 134
cattle, vii, viii, 1, 2, 16, 25, 147, 165
cellulose, 10, 15, 19, 49, 50, 54, 55, 57, 58, 62, 68, 69, 74, 142, 151, 155, 160, 164, 170
Central African Republic, 5, 110
cereal starches, 37
Chad, 5
challenges, 152, 158
chemical, 2, 8, 17, 20, 31, 38, 41, 69, 79, 131, 144, 151, 163, 168
chemical properties, 163
chemicals, 49, 68, 75
chicken, 4
China, 3, 5, 6, 9, 14, 15, 40, 50, 59, 61, 62, 66, 76, 77, 78, 79, 80, 154, 161, 162, 165, 170
cholesterol, 16
chromatography, 62, 130
chronic diseases, 130
City, 48, 51, 55, 60, 78
classes, 16, 130
classification, 4, 40
climate, 66, 110, 142, 152
climates, 50, 155
clone, 163
CO2, 57, 58, 60, 76
coal, 76, 162
coefficient of variation, ix, 109
coffee, 125
cogeneration, 163
collaboration, 156

College Station, 167
Colombia, 5, 109, 110, 124, 126
colonization, 12
color, 10, 16, 25, 28, 64
combustion, 160, 164
commercial, ix, 3, 23, 24, 25, 34, 43, 50, 59, 69, 72, 76, 156, 158, 161
communication, 72
competitiveness, 16
complex carbohydrates, 67, 72
complexity, 111
composition, 8, 11, 16, 17, 18, 20, 25, 36, 39, 40, 42, 43, 58, 61, 67, 68, 77, 79, 114, 138, 139, 144, 145, 159, 167
composting, 52
compounds, 14, 56, 62, 130, 135, 136, 137, 146
compression, 55
computation, 69
computer, 126
conditioning, 71
conference, 64
Congo, 110
Congress, 64, 126, 165
consensus, 165
construction, 53
consumers, 8, 13, 16, 23
consumption, vii, viii, 1, 2, 8, 12, 13, 16, 26, 33, 45, 55, 144, 153
containers, 76
contamination, 67, 149, 158
controlled studies, 154
convention, 42
conversion rate, 145
cooking, 14, 19, 20, 21, 26, 34, 35, 37, 39, 54
cooling, 54
copper, 73
correlation, 20, 21, 112, 121, 122
correlations, 20, 115
cost, vii, viii, 2, 3, 12, 16, 17, 25, 31, 32, 33, 46, 56, 59, 60, 65, 76, 142, 143, 147, 151, 159
cost effectiveness, 25
cost saving, 16
covering, 2, 24
crop, vii, viii, ix, x, 1, 2, 3, 5, 8, 10, 13, 15, 17, 18, 33, 46, 49, 50, 58, 65, 66, 75, 81, 109, 111, 114, 116, 123, 124, 129, 137, 141, 142, 143, 148, 150, 152, 155, 158, 160, 161, 165, 168
crop production, 81, 142, 161
crop residue, 15, 165
crops, viii, x, 2, 3, 46, 49, 50, 52, 57, 59, 63, 78, 81, 111, 114, 123, 124, 126, 141, 142, 143, 158, 160, 163, 165, 166, 168, 169, 170
crystalline, 151, 153

crystallinity, 164, 170
crystallization, 31
cultivars, ix, 4, 10, 22, 25, 49, 61, 79, 109, 111, 112, 113, 114, 119, 120, 121, 124, 125, 127, 142, 153, 159, 162
cultivation, vii, x, 3, 5, 6, 7, 8, 26, 46, 59, 60, 80, 123, 141
cultural practices, 155
culture, viii, ix, 21, 45, 56, 57, 58, 63, 72, 109, 110, 111, 114, 124, 133
culture medium, 56, 114
cysteine, 37, 146
cysteine-rich protein, 146
cytoskeleton, 111, 126
Czech Republic, 131

D

damaged grains, vii, 1, 2, 3, 12, 19, 21, 25, 42
database, 153
deconstruction, 151
decortication, 20
deficiencies, ix, 109
deficiency, 110, 111, 124
degradation, 20, 58, 69, 76
degree of crystallinity, 160
dehydration, 52
Department of Agriculture, 165
Department of Energy, 142, 161
depolarization, 111
deposition, 111, 126
depth, 13, 47
derived demand, 16
detection, 131, 136, 137
detoxification, 127
developed countries, vii, ix, 1, 3, 5, 15, 16
developed nations, 12
developing countries, 5, 16, 55, 142
diet, 16, 41, 130
diffusion, 40, 132, 135
digestibility, viii, 10, 17, 22, 26, 35, 36, 37, 39, 41, 43, 145, 146, 158
digestion, viii, 55, 76, 143
digestive enzymes, 139
diploid, 10, 142
diseases, x, 46, 129, 138, 158
displacement, 125, 126
distillation, 52, 143, 163
distilled water, 21
distribution, 8, 28, 40
diversification, 166
diversity, 124, 146, 155, 161, 164, 169
DNA, 5, 160, 170

DNA lesions, 160
DNAs, 11
DOI, 64
domestication, 15
dominance, 159
dosage, 21
dough, 14, 156
down-regulation, 146
drainage, 46
drought, vii, viii, ix, 1, 2, 5, 33, 45, 46, 47, 50, 64, 66, 129, 130, 142, 155, 158, 167
drought resistant cereal crop, vii, 1, 2
dry matter, 167
drying, 14, 47, 76, 152
DSC, 22

E

East Asia, 3, 110
economic competitiveness, 60
economics, viii, 2, 19, 23, 31, 81, 164, 165
editors, 138
effluent, 19
egg, 16, 17
Egypt, 3, 5, 9, 66, 78
El Salvador, 5, 14
electricity, 162, 163
elongation, 111, 112, 113, 124, 125
emission, 143, 162
encoding, 146, 163
endosperm, 4, 8, 10, 11, 23, 25, 34, 36, 37, 39, 40, 130, 145, 147, 164, 169
endothermic, 22
energy, vii, viii, 1, 14, 17, 19, 20, 33, 46, 50, 56, 60, 64, 65, 66, 67, 75, 76, 77, 78, 81, 138, 142, 144, 150, 151, 152, 155, 156, 158, 161, 163, 164, 165, 168, 170
energy density, 14
energy input, 60, 76, 142
engineering, 1, 42
entrapment, 58
environment, 2, 18, 20, 25, 167
environmental conditions, 11
environmental factors, 145, 159
environmental impact, 62, 165
environmental issues, 18
environmental protection, 78
enzyme, 21, 29, 33, 58, 69, 76, 146
enzymes, 10, 15, 21, 22, 23, 24, 29, 32, 33, 34, 37, 52, 54, 57, 58, 69, 71, 76, 145, 151
epidermis, 111
equipment, 53, 75, 76, 156, 158, 161
Eritrea, 3, 5

EST, 167
ester, 56
ethanol, viii, ix, 18, 19, 20, 21, 22, 26, 27, 29, 31, 34, 35, 38, 39, 42, 43, 45, 46, 49, 50, 51, 52, 53, 54, 55, 56, 57, 58, 59, 60, 61, 62, 63, 64, 65, 66, 67, 68, 69, 70, 71, 72, 76, 77, 78, 79, 80, 129, 132, 141, 142, 143, 144, 145, 146, 147, 148, 149, 151, 153, 154, 155, 159, 161, 162, 163, 164, 165, 166, 167, 168, 169, 170
ethyl acetate, 131, 133, 136
EU, 138
Europe, 3, 5, 8, 65, 76
evaporation, 32, 33, 51, 150
evidence, 37, 72, 121, 160
evolution, 57
experimental condition, 74, 121
experimental design, 114
exploitation, 26
exporter, 8
exposure, 125, 152
extraction, 25, 50, 143, 149, 153, 158
extracts, 63, 136, 138, 139
extrusion, 15, 20, 44

F

farmers, viii, 19, 26, 33, 60
fat, 11, 19
fatty acids, 72
feedstock, viii, x, 19, 43, 45, 46, 60, 63, 64, 68, 76, 79, 141, 142, 143, 144, 147, 149, 151, 155, 161, 166, 168, 169
fermentable carbohydrates, 49, 79
fermentation, viii, 3, 14, 15, 18, 19, 20, 21, 22, 23, 31, 42, 45, 51, 52, 54, 55, 56, 57, 58, 59, 61, 62, 63, 64, 67, 69, 71, 72, 74, 76, 77, 79, 80, 143, 149, 150, 151, 163, 164, 169
fermentation technology, viii, 45, 59
ferric ion, 138
fertility, 11, 126, 130, 142
fertilization, 48, 49, 61, 64, 66, 110, 156
fiber, 8, 10, 20, 21, 22, 50, 55, 66, 68, 69, 71, 76, 79, 163, 167
fiber content, 20
fibers, 10, 71
films, 25, 39
filtration, 29, 31, 32, 33
first generation, 142
fish, 15, 73
fish oil, 73
fitness, 159, 167
flavonoids, x, 129, 130, 134, 136
flavor, 15

flora, viii
flotation, 24
flour, vii, 1, 2, 13, 14, 15, 16, 19, 21, 25, 26, 27, 28, 29, 30, 34, 35, 37, 39, 40, 42, 131, 138, 146
flowers, 61
fluid, 20, 44
fluidized bed, 59
food, vii, viii, ix, x, 2, 3, 5, 8, 10, 12, 13, 14, 15, 16, 17, 23, 25, 33, 35, 36, 38, 39, 40, 41, 42, 43, 46, 50, 66, 129, 130, 131, 135, 138, 139, 141, 142, 145, 149, 154, 161, 164
food industry, 130
food production, 164
food security, viii, 2, 13, 33, 35
food spoilage, 130
forage crops, viii
force, 152
formation, 29, 37, 57, 110, 133, 160, 164
formula, 132, 133
fragments, 54
France, 3, 5, 8
free radicals, x, 129, 130, 132, 137
freezing, 76
frost, 152
fructose, 3, 23, 24, 31, 49, 62, 67, 73, 75, 148
fruits, 25, 63
functional food, 16
funding, 66, 160
fungi, x, 12, 17, 26, 38, 129, 131, 136
fungus, vii, 1, 3, 12, 26, 55, 58, 69
fusion, 10

G

gamete, 10, 145
gasification, 151
GDP, vii, 1, 3, 12
gelatinization temperature, 11, 33, 34, 54
genes, 11, 125, 146, 158, 159, 160, 164, 165, 166
genetic background, 159
genetic diversity, 64, 142, 160
genetic linkage, 167
genetics, ix, 18, 125, 156, 160, 167
genome, 142, 146, 160, 164, 166, 168, 170
genomics, 160
genotype, 11, 20, 22, 44, 158, 170
genotyping, 164, 170
genus, 4
Georgia, 153
Germany, 131
germination, 29, 47, 48, 61
germplasm, ix, 5, 142, 146, 148, 153, 156, 160

glucose, viii, 2, 3, 12, 15, 23, 24, 26, 27, 29, 31, 32, 35, 42, 43, 49, 52, 53, 54, 55, 56, 67, 68, 69, 71, 73, 74, 75, 77, 81, 143, 145, 147, 148, 151, 154
Glucose, vii, 1, 24, 27, 31, 32, 39, 40
glycerol, 52, 57, 73
governments, 60
grain yield, ix, 144, 154
grants, 160
granules, 10, 29, 34, 48, 145
grass, 4, 65
grasses, 124, 166
gravity, 59, 63, 67, 78, 79
Greece, 58, 66, 73, 77, 168
greenhouse, viii, 45, 66, 142, 152, 164
greenhouse gases, viii, 45, 142, 152
GRIN, 153, 161
growth, ix, x, 2, 12, 20, 46, 48, 49, 56, 57, 60, 61, 62, 63, 66, 72, 73, 74, 77, 81, 109, 110, 111, 121, 122, 123, 124, 126, 127, 129, 135, 136, 137, 142, 152, 158, 163, 167, 168
growth rate, 48, 60, 66
Guatemala, 5, 14
Guinea, 4, 5

H

Haiti, 5
harvesting, viii, 65, 75, 149, 158, 164, 167
health, 12, 15, 16, 26, 32, 50, 131, 147
health effects, 147
height, 66, 158, 159, 167
hemicellulose, 10, 54, 58, 62, 68, 69, 74, 151, 155, 170
hexane, 133, 136
history, 64, 169
homeostasis, 126
Honduras, 5, 14
human, 2, 5, 12, 13, 15, 16, 26, 38, 39, 49, 73, 130, 138, 144, 145, 153, 162
Human Food, vii, 1, 12
human health, 130, 138, 162
Hungary, 60, 66
hybrid, 4, 155, 156, 159, 161, 167, 169
hybridization, 142
hydrocarbons, 151
hydrogen, viii, 65, 66, 73, 74, 75, 77, 78, 80, 81, 149, 151, 163
hydrolysis, 20, 21, 22, 23, 27, 34, 35, 36, 38, 42, 50, 53, 54, 55, 57, 64, 69, 71, 79, 131, 139, 151, 160, 163, 164, 166, 169, 170
hydroxide, 23, 68, 69
hypothesis, 125, 126

I

ideal, 55, 60, 152
identification, 38, 110, 125, 160, 168
images, 22, 36
immobilization, 58, 63, 80
imports, viii, 8, 45
impregnation, 69
improvements, 159
impurities, 64
in vitro, 22, 35, 37, 39, 167
income, 8, 16, 59
India, viii, 1, 3, 4, 5, 6, 8, 9, 12, 13, 14, 16, 18, 23, 24, 25, 31, 33, 38, 40, 41, 42, 43, 45, 60, 63, 64, 66, 77, 154, 156, 161, 167
indirect measure, 22
Indonesia, 110, 169
induction, 111
industrial processing, 141
industries, vii, ix, 1, 3, 12, 15, 16, 18, 23, 129
industry, vii, viii, 2, 8, 12, 16, 18, 19, 23, 31, 38, 40, 52, 63, 147, 156, 161, 169
infancy, 156
infants, 14
infection, vii, x, 1, 3, 12, 25, 129
ingestion, 15
ingredients, 46
inheritance, ix
inhibition, 29, 39, 111, 125, 132, 134, 135, 136
inhibitor, 75
inoculation, 20, 57
inoculum, 21, 57
insects, 19, 37
insertion, 146
institutions, 156
integration, viii, 2, 29, 30, 163
integrity, 34, 36
interface, 29
internode, 158
investment, 53, 56, 59, 160
investments, 161
iodine, 11, 28, 34
ions, 110, 121, 122, 125
Iran, 60, 61, 62, 64, 153
irrigation, 47, 77, 158, 168
isolation, vii, 1, 2, 3, 12, 24, 26, 27, 41, 139
Israel, 8
issues, viii, 50, 65, 77
Italy, 3, 5, 8, 66, 76, 81, 154, 163, 164
Ivory Coast, 5

J

Japan, 8, 9, 14, 15, 16, 66, 78, 109, 127
Jordan, 8, 164, 167

K

K concentration, ix, 109, 122, 123
K^+, 121
Kenya, 3, 5, 14
ketones, 52
knowledge acquisition, 160
Korea, 5

L

laboratory studies, 76
lactic acid, 15, 23, 38, 44, 53, 57, 63, 75, 170
Laos, 110
Latin America, 5, 16, 126
leaching, 110
lead, viii, 45, 158
lesions, 160
life cycle, viii, 19, 65, 77
light, 10, 47, 64, 142
lignans, 139
lignin, 54, 68, 69, 79, 151, 155, 159, 160, 163, 164, 166, 167, 170
linoleic acid, 11, 56, 72
lipids, 10, 35, 56, 66, 77, 125, 148
liquid fuels, 77
livestock, viii, 2, 8, 12, 16, 25, 52, 147, 169
loci, 125, 146, 158, 166, 170
logging, 2, 46, 66
logistics, 143, 166
Louisiana, 153, 164
lower prices, viii, 2, 3, 12, 25
lying, 110
lysine, 10

M

magnitude, 31
maize, vii, viii, ix, x, 1, 2, 3, 8, 10, 11, 13, 14, 15, 16, 17, 18, 22, 23, 24, 25, 27, 35, 37, 39, 41, 43, 45, 46, 60, 62, 65, 109, 111, 112, 113, 114, 115, 116, 118, 119, 120, 121, 122, 123, 124, 125, 126, 141, 142, 143, 166
majority, 130, 145, 147
maltose, vii, 2, 3, 12, 23, 24, 26, 27, 29, 30, 31, 33, 49

maltose based products, vii, 2, 3, 12
management, 3, 12, 38, 40, 42, 152, 155, 156, 169
manipulation, 160
mannitol, 57
manufacturing, 46, 50
mapping, 125, 146
marketing, 143, 147, 155
Mars, 81
Marx, 169
mass, 57, 62, 72
mass loss, 57
mass spectrometry, 62
materials, 15, 31, 32, 33, 43, 54, 56, 67, 76, 118, 151, 163, 167
matrix, 10, 20, 21, 23, 29, 34, 36, 37, 145, 151
matter, 61, 67, 73, 154
Mauritania, 5
measurement, 111, 132
meat, 16, 25
media, 165
Mediterranean, 64, 161, 164
melting, 54
membrane permeability, 125
membranes, 135
metabisulfite, 37
metabolism, 16
metabolites, 133
metal ion, 37
metal ions, 37
meter, 23, 47, 48, 142
methanol, 10, 131
methodology, 27
methyl tertiary, 62
Mexico, 5, 6, 8, 9, 14, 15
Mg^{2+}, 121
microbial cells, 58
microbial communities, 73
micronutrients, 111
microorganism, 55, 58, 132, 133
microorganisms, 51, 57, 58, 73, 75, 131, 136, 137, 143, 151
microscopy, 22
microstructure, 21, 22
microstructures, 21, 22
Middle East, 3
Ministry of Education, 138
misconceptions, 143, 147
misuse, 18
mixing, 14, 132
models, 158, 163
modifications, 133
moisture, 2, 23, 26, 37, 47, 56, 57, 68, 71, 72, 75, 152

moisture content, 56, 57, 68, 71, 72, 75, 152
molasses, 17, 18, 59, 60, 142
mold, 12, 25, 38, 42, 147, 164, 167, 168
molds, 147
molecular biology, 125
molecular mass, 11
molecular structure, 163
molecular weight, 10, 25, 37
molecules, 151
Mongolia, 76, 79
monomers, 151
Morocco, 5
morphology, 8
Mozambique, 5
mucosa, 15
multiple regression, 121
multiple regression analysis, 121
multiplier, ix
multiplier effect, ix
mutant, 125, 146, 169, 170
mutation, 131, 133, 137, 139, 159
mutations, 130, 137, 142, 145, 146, 158, 159
mycotoxins, 12, 26, 29, 147

N

Na^+, 121
Namibia, 5
net energy balance, 60
Netherlands, 124
neutral, 127
Nicaragua, 5, 14
Nigeria, 4, 5, 6, 8, 9, 14, 24
nitrogen, 56, 59, 61, 63, 66, 67, 78, 142, 155, 164, 165
nonsense mutation, 168
North Africa, 3
North America, 3, 4
null, 146
nutrient, vii, ix, x, 10, 52, 109, 110, 111, 112, 114, 117, 118, 120, 121, 123, 124, 125, 126, 127, 138, 141
nutrients, ix, 46, 71, 109, 110, 111, 114, 121, 124, 155
nutrition, 12, 15, 131, 138

O

obstacles, 161
octane, 49
octane number, 49
OH, 110, 121, 124, 134

oil, 10, 14, 42, 56, 72, 80, 138, 139, 145, 151
Oklahoma, 162
oleic acid, 56, 72
omega-3, 73
operating costs, 56
operations, 147, 156
opportunities, 55
optimization, 38, 79, 161
organic matter, 61, 142
organs, 151
overlap, 159
ox, 49
oxidation, 71, 80
oxygen, 151

P

Pacific, 77, 166
Pakistan, 3, 5, 63
parallel, 133
parents, 159
pasta, 14, 15
pathogens, 12
penalties, 159
pepsin, 34, 35, 37
per capita income, 2, 25
percolation, 80
permission, 60
permit, 158
Peru, 5, 110
pests, 12, 43, 46, 158
petroleum, 45, 60, 143
pH, viii, 19, 21, 27, 28, 30, 45, 46, 58, 64, 71, 78, 110, 113, 114, 116, 117, 124, 127, 131, 133, 142
pharmaceutical, 3, 23
phenolic compounds, x, 16, 125, 129, 139
phenotype, 160, 168
Philippines, 48, 51, 55, 60
phosphate, 133, 156
phosphates, 110
phosphorus, 110, 124
photographs, 60
photosynthesis, 142, 150
physical properties, 20, 163
Physiological, 165
physiology, 125, 160
phytosterols, 24
pigs, 16
plant sterols, 16, 130
plants, ix, 4, 12, 14, 20, 47, 48, 50, 74, 78, 109, 110, 111, 121, 124, 125, 126, 143, 147, 151, 158, 160, 161, 163
plasma membrane, 111, 125

platform, 124
PM, 36
polar, 10
policy, 40
pollution, 17, 49, 56
polymer, 27, 29, 52, 160
polymerization, 20, 24, 37
polymerization process, 20, 37
polymers, 37, 68, 145
polymorphism, 5
polyphenols, 35, 38, 39, 130, 136, 138, 140
polyunsaturated fat, 73
population, viii, 2, 25, 47, 143, 146, 167, 169, 170
population density, 47
portfolio, 143
positive correlation, 112, 123
potassium, 23, 63
potato, 23, 50, 63, 64
poultry, 16, 17, 18, 42
power generation, 151
power plants, 162
precipitation, 110
preparation, 14, 25, 76
preservation, 67, 75, 150
prevention, 16, 138
probability, 120
producers, 12, 16, 17, 18, 19, 26, 60, 143, 147, 156
profitability, 31, 147
project, 31, 40, 138
protection, 37, 131
protein bonds, 21
protein components, 35
protein sequence, 15
protein structure, 20, 22, 37
proteins, 10, 11, 15, 22, 25, 26, 34, 35, 36, 37, 40, 68, 130, 146, 147, 168
proteolysis, 35
prototype, 158
Pseudomonas aeruginosa, x, 129, 131
purification, 139
purity, 24, 143, 163
pyrolysis, 167

Q

Queensland, 164

R

radicals, 134
rainfall, 2, 3, 18, 46, 110, 130
raw materials, 23, 31, 51, 59

reactions, 11
reactivity, 163
reagents, 23
recognition, 160
recommendations, 156, 157
recovery, 23, 24, 41, 56, 76
recrystallization, 54
reducing sugars, 30, 33
regions of the world, vii, x, 1, 2, 3, 141
regression, 120, 121
regression equation, 120, 121
regression line, 121
relatives, 146
relevance, 39
requirements, viii, 8, 45, 56, 160
research institutions, 60, 155, 158
researchers, 71, 73, 112
reserves, 45
residues, 19, 163
resistance, x, 2, 12, 39, 43, 46, 64, 66, 125, 126, 127, 129, 130, 131, 133, 137, 138, 147, 153, 158, 162, 164, 167, 168
resolution, 146, 164
resources, 45, 151, 160, 161
response, 16, 127, 168, 169
restrictions, 36
rheology, 39
risk, 67, 130, 152
Romania, 5
root, 2, 63, 111, 112, 113, 124, 125, 126, 127, 142
root growth, 111, 124, 126
root system, 2, 142
roots, 18, 47, 111, 120, 121, 122, 126, 127
routes, 3
rural areas, 13
rural development, 19
Russia, 3
Rwanda, 5

S

safety, 26
salinity, 46, 47, 50, 61, 66, 153, 162
salinity levels, 61, 162
salts, 59
savings, 60
scarcity, vii, 1, 2, 25
schizophrenia, 73
science, 138
scope, 56
second generation, 50, 66, 142
security, 13
sediment, 73

seed, 2, 4, 9, 10, 14, 43, 47, 60, 130, 142, 145, 146, 147, 155, 156, 159, 161, 169
seeding, 47
seedlings, 48, 111, 127
segregation, 162
senescence, 152
sensitivity, 125, 136, 158
sequencing, 160
shape, 4
shock, 29
shoot, ix, 109, 116, 120, 121, 122, 123
shoots, 116, 122, 123
showing, 6
Sierra Leone, 5
significance level, 113
silicon, 110
silk, 3
simultaneous saccharification, viii, 19, 42, 45, 54, 57, 63, 71, 79
Slovakia, 131
sludge, 74, 81
small intestine, 15
SO_4^{2-}, 110
society, 39
sodium, 23, 24, 37, 75
softwoods, 166
soil erosion, 142
soil type, 46
solid matrix, 56
solid state, viii, 45, 62, 64, 80, 149
solubility, 11, 22, 25
solution, ix, 11, 24, 71, 109, 110, 111, 112, 114, 121, 124, 127, 152
solvents, 42
Somalia, 3, 5
Sorghum bicolor, vii, viii, ix, x, 1, 2, 4, 5, 46, 61, 63, 65, 78, 79, 109, 131, 138, 141, 162, 164, 166, 167, 168, 169
sorghum cultivas, ix, 109
South Africa, 3, 5, 14, 34, 38, 40
South America, 3, 110, 111, 125, 141
Spain, 3, 5, 166
speciation, 126
species, ix, x, 3, 4, 12, 36, 58, 65, 72, 109, 111, 114, 120, 121, 123, 124, 125, 127, 141, 143, 147, 160, 165
Spring, 66
sprouting, 49
SSA, viii
stability, 158
stabilization, 80
stakeholders, ix

starch, vii, x, 1, 2, 3, 8, 10, 12, 15, 16, 18, 19, 20, 21, 22, 23, 24, 25, 26, 27, 28, 29, 30, 31, 32, 33, 34, 35, 36, 37, 38, 39, 41, 42, 43, 44, 49, 52, 53, 54, 57, 58, 61, 62, 64, 67, 74, 130, 141, 142, 143, 144, 145, 146, 149, 150, 151, 154, 160, 168, 169
starch crystals, 54
starch granules, 10, 21, 22, 23, 27, 29, 31, 34, 36, 145, 146
starch polysaccharides, 35
state, viii, 2, 8, 40, 56, 57, 61, 62, 71, 72, 77, 80, 136, 158
states, viii, 8, 19
stem sugar content, ix
sterile, 133
sterols, 125
storage, viii, 3, 10, 12, 14, 25, 51, 56, 65, 75, 136, 142, 143, 146, 149, 152, 156, 158, 164, 165, 166
stress, vii, ix, 2, 61, 109, 110, 112, 114, 115, 116, 117, 118, 121, 122, 123, 124, 125, 153, 155, 160
stress factors, 110, 114, 115, 123
structure, 10, 22, 25, 34, 38, 43, 68, 151, 159, 164
sub-Saharan Africa, viii
sub-Saharan Africa (SSA), viii
subsistence, 33
substitution, 13
substrate, 18, 19, 56, 57, 60, 62, 63, 73, 74
substrates, 18, 74
sucrose, 48, 49, 59, 67, 73, 75, 78, 147, 148, 154, 162
Sudan, 3, 4, 5, 6, 8, 9, 13, 14, 41, 167
sugar beet, viii, 45, 46, 50, 64, 166
sugar mills, 164
sugarcane, viii, 18, 45, 46, 49, 50, 59, 60, 63, 65, 66, 142, 148, 149, 158, 163, 166, 167
sulfate, 67
sulfur, 38, 73, 150, 163, 166
sulfur dioxide, 38, 150, 163, 166
sulfuric acid, 68, 71, 73
sulphur, 23
Sun, 38, 44
supply chain, 55
surface area, 36
susceptibility, 35, 54
sustainability, 66
sweat, 23
sweet sorghum, vii, viii, 4, 45, 46, 47, 48, 49, 50, 51, 52, 55, 56, 57, 58, 59, 60, 61, 62, 63, 64, 65, 66, 67, 68, 72, 75, 76, 77, 78, 79, 80, 81, 143, 145, 148, 149, 150, 153, 155, 156, 157, 158, 161, 162, 163, 164, 165, 166, 167, 168, 169, 170
swelling, 111
switchgrass, 65, 168
symptoms, 111
syndrome, 15, 126
synthesis, 127, 151, 164

T

Taiwan, 66, 78
tannins, x, 17, 36, 37, 129, 130, 135, 139, 140, 145, 146, 147
Tanzania, 3, 5, 6, 14
target, 144
techniques, viii, 45, 69, 75, 77, 78, 81, 111, 115, 124, 161, 162
technologies, viii, 15, 41, 43, 50, 65, 151, 161, 165, 166
technology, viii, ix, 2, 17, 24, 50, 59, 60, 63, 64, 67, 78, 126, 131, 143, 148, 150, 152, 169
temperature, 2, 11, 19, 23, 33, 34, 46, 49, 56, 57, 58, 64, 66, 68, 73, 110, 130, 131, 158
textbooks, 23
texture, 36, 39
Thailand, 3, 5, 66, 110
therapy, 16
thermal energy, 149
thermograms, 22
tissue, 10, 145, 151
Togo, 5
toluene, 133, 136
tones, 8
tooth, 47
total product, 8
toxicity, 50, 110, 111, 123, 124, 125, 126, 127
trade, 3, 8, 18, 62, 77
trade-off, 62, 77
traits, ix, x, 5, 141, 142, 145, 147, 158, 159, 161, 164, 167
transactions, 39
transesterification, 56, 73, 80
transformations, 62
transpiration, 2
transport, 3, 12, 25, 59, 111, 143
transportation, 66, 76, 149, 151, 152, 158, 168
treatment, 2, 14, 15, 23, 26, 28, 29, 34, 35, 37, 52, 68, 76, 112, 113, 116, 117, 120, 121, 138, 167
trial, 164
triploid, 10, 145
tryptophan, 10
tunneling, 48

U

UK, 41, 62, 63, 163
ultrasound, viii, 2, 24, 27, 28, 29, 30, 42

UNESCO, 41
uniform, 47
United, vii, 1, 3, 4, 5, 6, 8, 9, 12, 50, 78, 165, 166
United States, vii, 1, 3, 4, 5, 6, 8, 9, 12, 50, 78, 165, 166
urban, 13
urbanization, 2, 13, 25
Uruguay, 5
U.S.A., 3, 5, 8, 14, 15, 24, 39, 40, 41, 43, 62, 63, 64, 65, 66, 146, 164
USDA, 64, 66, 153
Utilization of the sorghum, vii, 1, 12
UV, 133
UV light, 133

V

vacuum, 51
Valencia, 166
variables, 57, 76
variations, 13, 139, 160
varieties, vii, viii, x, 4, 5, 11, 15, 17, 18, 19, 21, 34, 43, 46, 61, 66, 68, 74, 78, 130, 141, 145, 147, 148, 150, 152, 153, 154, 155, 156, 158, 160, 161, 167
vegetables, 15, 130, 139
vehicles, 49
Venezuela, 3, 5
versatility, 160
Vietnam, 110
vinasse, 52
viscosity, 14, 22, 34, 54, 73, 143, 146
vitamin C, 130
vitamins, vii, 1, 19, 33
vulnerability, 55

W

war, x, 129

Washington, 165
waste, 23, 71, 80
waste water, 23
water, viii, 2, 5, 11, 20, 23, 25, 28, 32, 33, 45, 46, 47, 49, 50, 53, 54, 56, 57, 60, 62, 66, 68, 69, 71, 72, 73, 78, 111, 132, 142, 143, 145, 150, 154, 155, 161, 164
water quality, 49
water vapor, 25
wave propagation, 29
web, 20, 22, 37
weight gain, 147
weight loss, 58
West Africa, 3, 13, 14
wild type, 145, 159
Wisconsin, 66
wood, 71, 80, 164, 166
workers, 4, 59
worldwide, 2, 15

Y

yeast, viii, 19, 21, 45, 51, 52, 56, 57, 58, 59, 61, 62, 63, 64, 67, 69, 71, 72, 75, 77, 80, 131, 136, 143, 148, 149, 151
Yemen, 4, 5
yield, vii, ix, x, 2, 6, 7, 8, 19, 20, 21, 22, 23, 29, 42, 46, 47, 48, 49, 50, 54, 56, 57, 58, 59, 60, 61, 63, 64, 66, 67, 68, 69, 70, 71, 72, 73, 74, 77, 78, 79, 111, 127, 141, 142, 144, 147, 148, 154, 155, 158, 159, 160, 162, 163, 165, 166, 167, 169, 170
yolk, 17

Z

Zimbabwe, 5